Time-Resolved Spectroscopy in Complex Liquids
An Experimental Perspective

Renato Torre
Editor

Time-Resolved Spectroscopy in Complex Liquids
An Experimental Perspective

 Springer

Renato Torre
European Lab for Non-linear Spectroscopy (LENS) and Dip. di Fisica,
Polo Scientifico, University of Firenze,
Via Nello Carrara 1,
I-50019 Sesto Fiorentino, Firenze,
Italy

Time-Resolved Spectroscopy in Complex Liquids

Consulting Editor: D. R. Vij
Kurukshetra University
E-5 University Campus
Kurukshetra 136119
India

Library of Congress Control Number: 2007929383

ISBN 978-0-387-25557-6   e-ISBN 978-0-387-25558-3

Printed on acid-free paper.

Printed on acid-free paper.

9 8 7 6 5 4 3 2 1

springer.com

"Nel mezzo del cammin di nostra vita mi ritrovai per una selva oscura, ché la diritta via era smarrita."

"In the middle of my life's jorney I woke to find myself in a dark wood, where the right road was wholly lost and gone".

Dante Alighieri
(Firenze, 1265)

Contents

Contributing Authors

Bartolini, Paolo
European Lab. for Non-Linear Spectroscopy (LENS), Universitá di Firenze, via N. Carrara 1,
I-50019 Sesto Fiorentino, Firenze, Italy
INFM-CRS-SOFT c/o Univ. la Sapienza, Roma, Italy
bart@lens.unifi.it

Čopič, Martin
Faculty of Mathematics and Physics, University of Ljubljana, Slovenia
J. Stefan Institute, Jamova 39, Ljubljana, Slovenia
martin.copic@fmf.uni-lj.si

Drevenšek Olenik, Irena
Faculty of Mathematics and Physics, University of Ljubljana, Slovenia
J. Stefan Institute, Jamova 39, Ljubljana, Slovenia
irena.drevensek@ijs.si

Eramo, Roberto
European Lab. for Non-Linear Spectroscopy (LENS), Universitá di Firenze, via N. Carrara 1,
I-50019 Sesto Fiorentino, Firenze, Italy
INFM-CRS-SOFT c/o Univ. la Sapienza, Roma, Italy
eramo@fi.infn.it

Li, Yun-Liang
Departments of Chemistry and Physics, Institute of Optical Sciences, University of Toronto, 60
St. George St., Toronto, Ontario, Canada
li@lphys.chem.utoronto.ca

Manzo, Carlo
CNR-INFM Coherentia, Universitá Federico II c/o Dip. Scienze Fisiche, Complesso Univ. Monte
S. Angelo, via Cintia, I-80126 Napoli, Italy
carlo.manzo@na.infn.it

Marucci, Lorenzo

CNR-INFM Coherentia, Universitá Federico II c/o Dip. Scienze Fisiche, Complesso Univ. Monte S. Angelo, via Cintia, I-80126 Napoli, Italy
lorenzo.marrucci@na.infn.it

Miller, R. J. Dwayne

Departments of Chemistry and Physics, Institute of Optical Sciences, University of Toronto, 60 St. George St., Toronto, Ontario, Canada
dmiller@lphys.chem.utoronto.ca

Milne, Christopher J.

Department of Chemistry, University of Toronto, Canada
Present Address: Swiss Light Source, Paul Scherrer Institut, CH-5232 Villigen, Switzerland
chris.milne@utoronto.ca

Paparo, Domenico

CNR-INFM Coherentia, Universitá Federico II c/o Dip. Scienze Fisiche, Complesso Univ. Monte S. Angelo, via Cintia, I-80126 Napoli, Italy
paparo@na.infn.it

Taschin, Andrea

European Lab. for Non-Linear Spectroscopy (LENS), Universitá di Firenze, via N. Carrara 1, I-50019 Sesto Fiorentino, Firenze, Italy
INFM-CRS-SOFT c/o Univ. la Sapienza, Roma, Italy
taschin@lens.unifi.it

Torre, Renato

European Lab. for Non-Linear Spectroscopy (LENS) and Dip. di Fisica, Universitá di Firenze, via N. Carrara 1, I-50019 Sesto Fiorentino, Firenze, Italy
INFM-CRS-SOFT c/o Univ. la Sapienza, Roma, Italy
torre@lens.unifi.it

Vilfan, Mojca

J. Stefan Institute Jamova 39, Ljubljana, Slovenia
mojca.vilfan@ijs.si

Preface

The investigation of liquid properties is as old as the human desire to observe natural phenomena. The paramount relevance of such state of matter is obvious, spanning from the basic theoretical model of random systems to the more advanced technical applications. Nevertheless the interpretation and understanding of liquid properties remains a challenge in materials science.

In the liquid state, the potential and kinetic energies are characterized by similar values, differently from the gas or solid/crystal phases. This fundamental physical property prevents a description of the liquid structure independently from its dynamics. The intermolecular forces and molecular motions are strictly interconnected, producing the peculiar features of liquid phases.

In particular when the intermolecular potential is characterized by specific features, as anisotropic and/or long-range interactions, the liquid state shows local dynamic aggregation and structuring phenomena. This is the distinguishing characteristic of *complex liquids*. Typical examples are the hydrogen-bonded liquids, glass formers, polymers and liquid crystals. Also relatively simple molecular liquids, e.g., CS_2 and benzene, clearly show complex dynamics, evidence of local structuring effects.

During the last years, researchers have undertaken a steady effort to improve the knowledge of liquid matter, both from experimental and theoretical points of view. New spectroscopic techniques, based on nonlinear optical phenomena, have been realized and applied to the study of simple molecular liquids. On the other hand, standard experiments have been applied to liquids characterized by particularly complex dynamic processes.

In this book, we collect a series of chapters dedicated to the state-of-art studies of optical spectroscopy in the time domain on complex liquids of different nature. In Chap. 1, a new nonlinear spectroscopic technique, 2D-Raman, is comprehensively reviewed. This is probably the most promising experimental tool able to collect truly new information on fast dynamics in liquid matter. Chapter 2 is dedicated to optical Kerr effect techniques and to the investigation of relaxation dynamics in complex liquids, inclusive of relative slow collective phenomena. Transient grating spectroscopy is reviewed in Chap. 3, introducing the recent experimental improvements and their application to the study of viscoelastic phenomena in glass formers. Chapter 4 describes the dynamics of confined liquid crystals as measured by dynamic light scattering experiments. In Chap. 5, the host–guest interactions of dye molecules in liquid-crystal matter is analyzed by time-resolved fluorescence and dichroism spectroscopy.

Firenze

RENATO TORRE

Chapter 1

TWO DIMENSIONAL FIFTH-ORDER RAMAN SPECTROSCOPY

A New Tool for the Study of Liquid State Dynamics

Christopher J. Milne, Yun-Liang Li, and R. J. Dwayne Miller

Abstract   Fifth-order Raman spectroscopy has the potential to enable chemists to directly probe the intermolecular potential between molecules within the liquid state. Understanding the liquid state has been a long-standing goal in physical chemistry as most chemical and biological processes occur in solution. This new spectroscopy has the capacity to examine these low-frequency intermolecular modes and gives direct access to the many-body potential of liquids. Effectively this spectroscopy provides a direct window on the anharmonic motions of molecules that distinguish the dynamics of the liquid state. The observable in this experiment provides an unforgiving test of potentials used to model liquids and as such is an important new tool for arriving at a first-principles treatment of liquids. The experiment is complicated by the extremely small signals associated with the Raman processes probing the intermolecular frequency correlations. A new approach based on diffractive optics has solved the last remaining obstacles to the successful experimental implementation of this spectroscopy. This chapter presents a thorough review of the field to date, covering both the novel theoretical and experimental developments that have proven necessary to achieving this success. The heterodyned fifth-order Raman response of liquid CS_2 and liquid benzene has been measured and characterized by specifically exploiting the passive-phase stabilization provided through the use of diffractive optics. The dynamics of the two liquids is compared with each other and various recent theoretical results. The measurement of the low-frequency Raman two-time delay correlation function indicates the intermolecular modes of both liquids to be primarily homogeneously broadened and that the liquid loses its nuclear rephasing ability very rapidly. This rapid loss of nuclear correlations indicates a lack of modal character in the low-frequency motions of liquid CS_2 and liquid benzene. The chapter concludes with the authors' vision for the future of the experiment and provides a roadmap toward achieving the ultimate goal of measuring the fifth-order Raman response of liquid water.

1.1   Introduction

This book is dedicated to the application of various linear and nonlinear optical spectroscopies to the study of complex liquids. The importance of this subject is readily justified as a significant fraction of chemistry, and all of biology can be considered to involve such systems. The relative motions and correlations between molecules in liquids are a direct consequence of the intermolecular potential, the anharmonic terms of which can be related to the fluctuation and dissipation processes that govern the physical behavior of the liquid as well as dictate the course of reactions. In complex liquids, there are a plurality of contributions to the intermolecular potential that lead to stronger attractive forces and greater correlations than in unstructured liquids. In all cases, the liquid state is a marginally stable phase in which the attractive forces between molecules are sufficient to produce a condensed phase but not so strong as to lead to solid or glass state formation. Molecules are free to diffuse relative to one another in the liquid state and it is possible to attain high densities of reactive species in solution phase. Both effects greatly increase collision frequencies above alternative media and accounts for the fact that most chemistry is explored using solution phase processing. The advantage of complex liquids, as defined here, is that the additional intermolecular forces such as hydrogen bonding and electrostatic terms can act to further lower barriers in reaction complexes and, in the extreme case of biological systems, resulting in preferred orientation to the point of molecular self-assembly. In this context, the study of complex liquids and the degree of dynamic correlations is an important prelude to understanding the more structured environment in biological systems.

The study of liquids essentially reduces to the study of the time-dependent loss in correlations in instantaneous structure that allows the system to take on liquid properties. It is known from X-ray and neutron scattering that the structure of liquids is comprised of a radially disordered distribution of molecules or atoms in a shell-like structure in which the reduced density function typically only shows short-range order extending over 1–2 shells [1]. The reduced density functions of liquids are therefore significantly different from those of solids that show long-range order. In contrast, the structure of liquids and glasses are very similar. The only thing that separates the two classifications of matter is the timescale or dynamics of the various length scales of motion. The most distinguishing, single, bulk parameter between liquids and glasses is that liquids do not support shear modes [1]. Transverse motions, even with a forcing function, are strongly damped in the liquid state and do not propagate. This distinction is often used as one of the central definitions of the liquid state. This definition is only strictly true for weakly associated atomic liquids held together principally by van der Waals forces, more complex liquids have several contributions to the intermolecular potential with different length scale dependencies to the attractive terms in the potential. Dipolar and higher order terms, hydrogen bonding, and electrostatic interactions can lead to significant short-range intermolecular forces that become comparable to covalent bonds in aggregate and certainly much greater than kT. Thus, these forces lead to correlations between molecules over a range of different length scales as they undergo stochastic fluctuations. In this context, the atomic or molecular pair correlation functions of liquids represent time and configurationally averaged structures. There are a myriad of different nuclear configurations within a liquid, all interconverting and extensively sampling this structural phase space. Ideally, one would like to observe molecular motions over the complete range of time and length scales to fully understand how molecules move relative to one another in the liquid state. At short times, there would be long range correlations that could be described at least qualitatively within a modal description and then on longer times the motions would lose correlations and the system would cross over to Brownian or diffusive uncorrelated motions. This distinction is shown schematically in Fig. 1.1 for a typical spectral density of intermolecular frequencies of motion. These liquid state motions would be extremely anharmonic relative to bound

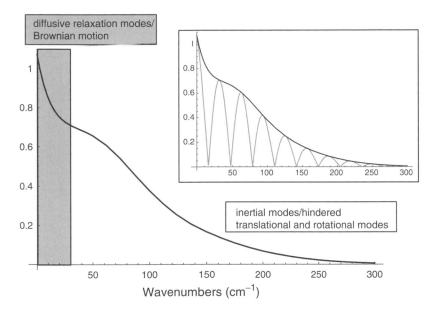

Fig. 1.1 Typical spectral density of states for liquids covering the intermolecular region of the spectrum. The low-frequency *shaded area* represents the region of the spectrum corresponding to diffusive relaxation processes, as the energy increases we move into nondiffusive inertial-type motions. The *inset* represents a microscopic examination of the same spectral lineshape in which the spectrum is constructed of a distribution of nuclear configurations giving rise to much narrower lineshapes within an inhomogeneous lineshape vs. the homogeneous limit in which these configurations rapidly interconvert

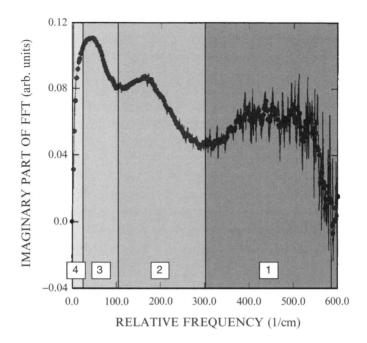

Fig. 1.2 The experimentally determined spectral density of states of liquid water using the optical Kerr effect ($\chi^{(3)}$-linear response) is shown for comparison [3]. Water is one of the most structured liquids in which there is large separation in timescales. The frequency range between 300 and 1,000 cm^{-1} (**1**) is related to librational motions (note that the spectrum represented here is artificially truncated at 600 cm^{-1} due to laser bandwidth limitations), the peak at 170 cm^{-1} (**2**) is hindered translational motion of the heavy O atoms, the 60 cm^{-1} mode (**3**) is transverse or shear motion, and below 25 cm^{-1} (**4**) corresponds to diffusive relaxation and hydrogen bond breaking [8]. The issue of inhomogeneous to homogeneous interpretations of the dynamic structure of liquid water still holds despite this additional structure. The agreement is quite good with the theoretical calculation of the corresponding spectral density of states for water but the calculations are rather insensitive to basis with respect to this observable. Reprinted with permission from [3]. Copyright 1994, American Chemical Society

states of matter and would involve a chaotic sea of density fluctuations with correlated and anticorrelated motions leading to void space formation to assist molecular diffusion. Could we ever experimentally capture such a microscopic picture of liquids?

The above question pertains to a direct experimental determination of the liquid's many-body potential. In principle, information on the intermolecular potential is contained within the dynamic lineshapes as part of the spectroscopic observable. One has to then consider the appropriate frequency range for the dynamical variables of interest. For liquids, intermolecular and atomic motions have Fourier components typically less than 200 cm^{-1} (see Fig. 1.1). Only for more structured liquids such as water does the frequency range extends up to

$1,000 \text{ cm}^{-1}$ (see Fig. 1.2) [3]. The relevant frequency range in descending order includes hindered rotational motions at the high frequency cut off describing the particular fluid's motions, to hindered translations, strongly damped transverse or shear motions out to density fluctuations. In this frequency range, experimental approaches need to employ either GHz to THz sources to study dipolar responses or low frequency Raman scattering. On timescales shorter than the onset of diffusive or Brownian motion ($t < 1$–10 ps typically), it is convenient to think about the collective correlated motions of liquid molecules as modes in analogy to solid state phonons [4]. Within this basis, the many-body potential could be directly determined from experiment if it were possible to map out the relaxation pathways interconnecting energy exchange between the different types of motion. For example, using the assumption that the linewidth of a particular mode within a modal basis is solely due to lifetime ($T1$) broadening, the spectrum could be directly inverted to the liquid many-body potential.

A cursory inspection of Figs. 1.1 and 1.2 shows the problem with such an approach: there are no well-separated transitions with discernible linewidths as one would observe for electronic transitions or intramolecular vibrations. The spectra of liquids in the range characterizing their interatomic/intermolecular motions are very broad and featureless. This aspect of liquids is to be expected as liquid structure represents a nuclear continuum. In fact, it is this very feature that drives reaction dynamics in solution phase; reactions in solution phase are coupled to the nuclear continuum of states that leads to wavefunction collapse on the product surface [5]. There are no quantum recurrences within a single crossing event and one describes such events using the language of dynamics, as opposed to time-dependent eigenstates. The upper limit to barrier crossing involving a solvent coordinate is determined by the various types of liquid motions being discussed herein. For barrierless reactions, the fastest time constant is determined by the type of liquid motions and the coupling coefficient to the reaction coordinate [6]. Other reaction rates are similarly affected by the barrier crossing dynamics, weighted by the probability of statistically sampling the barrier crossing region, as described within the context of transition state theory [7]. This relationship further emphasizes the importance of understanding liquid dynamics. However, it is impossible to apply the standard approaches in linear spectroscopy (single time or frequency variables) to connect lineshapes to dynamical processes. Important information can still be determined using linear spectroscopies with respect to the long timescale relaxation processes and spectral density of states as discussed throughout this book and this information can be used to refine liquid state theory. The primary objective of this chapter is to outline an experimental approach to *directly observe the anharmonic motions* in liquids that are intrinsic to this state of matter. Linear spectroscopy is not equal to the task, some form of multidimensional spectroscopy is required to dissect the broad featureless liquid spectra.

Interpreting the lineshape representative of liquid motions shown in Fig. 1.1 is a classic example of homogeneous vs. inhomogeneous broadening issues surrounding any non-Lorentzian lineshape [8]. This problem was first addressed in NMR spectroscopy for congested spectra and led to the development of the spin-echo technique that is the precursor of all higher order pulse sequences used in current multidimensional spectroscopy. Analogues of the NMR spin echo have been reported based on electronic transitions (photon echo) [9], Raman transitions (Raman echo) [10, 11], and vibrational transitions (IR echo) [12], with more complex pulse sequences similar to NMR to surely follow. All these two-dimensional spectroscopies share a common feature of eliminating the effects of inhomogenous broadening from the lineshape. The experiment consists of a two pulse sequence in which the first pulse sets up a coherence between the ground and excited states of the specific transition in question, which drives a material polarization. This polarization in turn generates a signal field that undergoes a free induction decay that is simply the inverse of the lineshape [13]. The various excited transitions may still be coherently related to the initial excitation, however it is impossible to determine as the macroscopic polarization decays due to the frequency distribution and associated dephasing. In this regard, monitoring the free induction decay in the time domain contains no new information beyond that found in the broad frequency domain spectrum; this signal field is simply the Fourier transform of the frequency domain spectrum. In multidimensional spectroscopic approaches this initial preparation pulse is followed by a second pulse that creates a new coherence in which the initial polarization has undergone a $180°$ phase shift. The phase of the high-frequency transitions within the lineshape contributing to the material polarization is now retarded relative to the slower evolving low-frequency components. All the different frequencies within the inhomogeneously broadened lineshape will exactly rephase at exactly the same time interval as that separating the initial pulse and the rephasing pulse. This refocusing of the macroscopic polarization is referred to as the "echo." Scanning the time delay and measuring the spectrum of the emitted signal field from this coherence generates the 2D spectrum of interest that casts out the homogeneous and inhomogeneous contributions to the lineshape. In addition, the off-diagonal features in the 2D spectrum give directly the anharmonic couplings between the nonstationary modes. It is exactly this kind of information that one requires to construct the many-body potential of liquids. In fact, the rephasing process itself can be viewed as effectively a measurement of the timescale over which the different types of excited motions are correlated. In other words, 2D measurements can in principle directly access the memory function of liquids as well as the anharmonic couplings between the different frequency components of the liquid spectrum, providing a direct view of the anharmonic motions in liquids.

The experimental challenge has been to come up with the proper pulse sequences to generate the necessary coherences to probe the bath dynamics. As discussed above, to excite liquid modes requires driving intermolecular motions with high frequency cut offs in the 100–$1,000\,cm^{-1}$ frequency range. To drive intermolecular motions through dipolar couplings to the electric field in this range requires THz pulses. The peak power and pulse duration are insufficient with current THz technology. At the time of writing, a nonlinear THz experiment has yet to be successfully conducted. In addition, there is already ample evidence from theoretical calculations and study of solvation dynamics that liquid modes in the THz frequency range are strongly damped [14]. The experiment would require single cycle pulses to attain sufficient time resolution to view the rephasing processes and even in this limit the time/frequency resolution may not be sufficient to access the anharmonic features in the motions of interest. The only other means of exciting low-frequency motions in the range relevant to liquid dynamics is to use a Raman process. The differential change in polarizability with displacement behind the Raman process is also a unique probe of the liquid dynamics through the effect of collision-induced changes in the polarizability. In this experimental approach, the laser pulse must contain sufficient spectral bandwidth to drive the frequency components of interest. With current laser technology, it is not a problem to generate pulses as short as a few femtoseconds in which there is more than sufficient bandwidth to drive all possible liquid phase motions. The excitation process can truly be made in the impulsive limit and there is sufficient time resolution to follow motions that are strongly damped and capture anharmonic components of even less than half a period. The peak power of these sources is enormous, well sufficient to efficiently drive Raman processes all the way up to the dielectric breakdown threshold of the material. It is really the dielectric breakdown threshold that defines the upper limit to how strongly driven these modes can be.

The prospect of directly accessing information on liquid modes using echo-based spectroscopic methods is an intriguing concept. We would essentially be directly probing the bath in which chemical reactions take place. Whatever the mechanism of driving the liquid modes, there must be a rephasing pulse sequence. The very notion that it may be possible to intervene in a system as marginally stable as a liquid, that is undergoing large amplitude stochastic fluctuations, and refocus the motions contradicts our sense of the liquid state. This concept of a "bath echo" is simply further recognition that on short enough timescales liquids have motions that are largely inertial and can be described using a modal basis. We will lose our ability to refocus the bath motions once the system of driven modes has irreversibly undergone an inelastic collision/energy exchange or a dephasing process. The timescales and frequency ranges involved in such processes provide a much more microscopic basis from which to model the liquid state. The overarching hope here is that such experiments will enable direct access to the highly anharmonic motions of the liquid state.

The original experimental concept of performing such bath echo experiments was based on the *optical Kerr effect* (OKE) to attempt to use the anisotropic polarizability to enable polarization selection of the signal [15]. OKE has been extensively used to study low-frequency motions of liquids (see Chap. 2). The experiments that were ongoing at the time led Tanimura and Mukamel to theoretically explore the nonlinear response function of liquids using a harmonic Brownian oscillator basis for the bath [16]. In a real tour de force of theory, they discovered that there was a well-defined rephasing process at the fifth order of the nonlinear interaction with the laser fields. This rephasing pathway is very distinctive and the most novel feature of this new form of spectroscopy. Conventional echoes (spin echo, IR echo, photon echo) are four-wave mixing experiments or third order with respect to input fields ($\chi^{(3)}$ nonlinear response). One field (first excitation pulse) creates the initial coherence and then there is a two-field interaction that rephases the coherence by cycling the excited state coherence through the ground state to introduce a $180°$ phase change in the polarization. The emitted signal field is then detected to constitute the fourth field. A strictly analogous Raman process would require eight fields and is formally a $\chi^{(7)}$ or seventh-order nonlinear response. The Raman excitation process requires two fields at each step such that there would be a two-field interaction to initialize the coherence, followed by a four-field interaction to cycle the coherence, and finally a seventh field would be required to stimulate the Raman scattering field that would constitute the eighth field in the material interaction [17]. The theoretical work of Tanimura and Mukamel discovered that it is possible to generate rephasing pathways at the fifth-order or $\chi^{(5)}$ nonlinear response level by involving a two-quantum transition. Rather than interact with the same levels as done conventionally, the introduction of a coherence involving a Raman overtone provides a unique rephasing pathway as shown schematically in Fig. 1.3. There are two inherent assumptions in this model: (1) the bath can be approximated as harmonic to ensure that the overtone is twice the frequency of the fundamental and (2) dephasing is level independent, i.e., the dephasing of the overtone is correlated to the same bath motions and to the same degree as the fundamental. As discussed, liquids are expected to be highly anharmonic and marginally stable such that the overtone of a liquid mode is likely to not even be a bound state. The original basis for this theoretical development is only marginally valid at short times. However, this work laid the foundation for a flurry of theoretical activity that has now shown that $\chi^{(5)}$ studies of liquid dynamics do indeed cast out the anharmonic motions of greatest interest and represents one of the most rigorous tests for theoretical models of the liquid state.

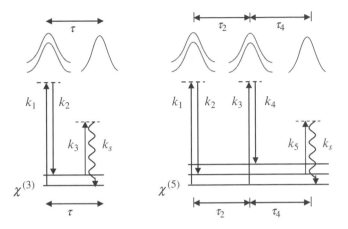

Fig. 1.3 Pulse sequence (*top*) and energy level diagram (*bottom*) for third-order (*left*) and fifth-order (*right*) Raman spectroscopy. The nonresonant third-order spectroscopy consists of a pair of time-coincident pump pulses ($\mathbf{k_1}$, $\mathbf{k_2}$) followed by a probe pulse ($\mathbf{k_3}$) after a time delay τ. The fifth-order nonresonant spectroscopy consists of two pump pulse pairs ($\mathbf{k_1}$, $\mathbf{k_2}$, and $\mathbf{k_3}$, $\mathbf{k_4}$) separated by a time delay τ_2, followed by a probe pulse ($\mathbf{k_5}$) after a second time delay τ_4. In both cases a signal field ($\mathbf{k_s}$) is generated by the scattering of the probe off the pump-induced grating in the sample

The proposed $\chi^{(5)}$ experiments by Tanimura and Mukamel stimulated a considerable experimental effort to observe the bath echo using the predicted $\chi^{(5)}$ nuclear response function. Initial attempts were made to resolve the fifth-order signal in liquid CS_2, the Raman community's sample of choice [58–60]. The results were contrary to the predicted response function, and yet the experiments themselves had been performed by three independent groups, all with consistent results. The unexpected answer was arrived at several years later when it became evident that the signal measured was not the desired fifth-order signal, but was instead a contaminant third-order signal that had not been accounted for in the original experimental design [18–20]. The main problem in this class of experiments is that the rephasing step involves a two-quantum or Raman overtone transition. The one-quantum Raman transition is already a weak transition but the two-quantum transition will be considerably weaker. Lower order cascaded $\chi^{(3)}$ responses can easily dominate the fifth-order signal of interest if care is not taken. New experimental methodologies were needed, more akin to NMR methods, in which phase sensitive detection with a reference field could be used both for phase discrimination and signal amplification. New beam geometries for better phase matching discrimination between the fifth-order and cascaded processes were also needed in which strict adherence to phase matching could be conserved to avoid parameter space in which the much larger cascaded signals would dominate and inadvertently bias the alignment procedure. This objective was the driving force for the development of diffractive optics-based nonlinear spectroscopy [11, 21, 22] in which the desired beam geometry as well as a reference field for a local oscillator for heterodyned phase sensitive detection could be designed into a single optic. The diffractive optic approach provides all the necessary beams in a passively phase-locked condition that was critical to increasing the signal-to-noise ratio by over two orders of magnitude over conventional detection of the signal intensities. The real power of diffractive optics-based nonlinear spectroscopy was fully demonstrated with the first full characterization of the true fifth-order signal for liquid CS_2 in 2002 [23]. This general methodology represents a significant advance to nonlinear optics in general and will undoubtedly form the basis for extending these studies to other liquids.

It is now established that this form of six-wave mixing spectroscopy gives a direct probe of the anharmonic motions in liquids and holds the greatest promise for providing a direct experimental determination of the many-body potential of the liquid state. This chapter expands on the recent theoretical and experimental advances that have contributed to our current understanding of the fifth-order Raman experiment. The inherently low-signal aspects of this experiment represent an enormous experimental challenge to extend this form of spectroscopy to other systems; likewise the complex source terms and high sensitivity of the signal to anharmonicity in the intermolecular potential and nonlinear terms in the electronic polarizability make this experiment equally challenging for the theoretical community. The new information this experiment accesses, however, provides strong motivation for rejuvenating efforts in this direction. This chapter discusses the specific requirements for extending this spectroscopy as well as provide a road map for what lies ahead.

1.2   Theory

One of the initial time domain Raman experiments enabled by the advent of coherent femtosecond laser pulses is the optical Kerr effect, third-order Raman spectroscopy [24]. This spectroscopy gives insight into the Raman excitation process as a first step toward 2D methodologies. In this spectroscopy, two time-coincident pump pulses induce a Raman coherence between two low-frequency modes in the liquid sample. After some time delay a probe pulse scatters off this coherence, generating the signal in the phase-matching direction $k_s = (k_1 - k_2) + k_3$ (see Fig. 1.3). In 1985, Loring and Mukamel showed the equivalence between the ultrafast time domain response and the frequency domain light-scattering experiment: the third-order nonlinear spectroscopy contains no additional information about the low-frequency modes in the liquid [25, 26]. It is not possible to distinguish homogeneous from inhomogeneous contributions to the lineshape from this single time variable spectroscopy. To access a second time variable, to refocus the polarization and access the time evolution of the bath directly, one needs a second pair of pulses as shown schematically in Fig. 1.3 with the various time orderings labeled. This second pulse pair transfers the Raman coherence to a superposition state involving a two-quantum transition. Within a harmonic oscillator approximation, the induced polarization between the first and second excited states of the bath modes have exactly the same frequency distribution but there is a sign change in the polarization between the 0–1 and 1–2 coherences. This effect gives the $180°$ phase shift in the sample polarization for these field interactions that is essential for refocusing the inhomogeneous contributions to the lineshapes. The time delay between the two excitation pulse pairs is labeled τ_2 (T_1 is also used in the literature). The induced polarization will not radiate within the same spectral range; a fifth pulse is needed to undergo stimulated Raman scattering to probe the bath polarization. The signal field that emerges represents the sixth field; the five input fields make this induced polarization a fifth-order interaction characterized by the $\chi^{(5)}$ nuclear susceptibility. The relative time delay between this probe pulse and the first excitation pulse pair is labeled τ_4 (T_2 is also used in the literature). In the case of an inhomogeneously broadened system the second coherence will cause a rephasing of the first coherence when $\tau_2 = \tau_4$, resulting in an echo-like signal analogous to photon-echo spectroscopy. The signal generated meets the phase matching requirements $\mathbf{k_s} = (\mathbf{k_1} - \mathbf{k_2}) - (\mathbf{k_3} - \mathbf{k_4}) + \mathbf{k_5}$; in a noncollinear phase-matching geometry this results in a unique signal direction making spatial isolation possible. The framework surrounding nonlinear response functions is well established [27] and is not covered here, we instead focus on the relevant results and their analysis. When an electromagnetic field interacts with a medium, it induces a polarization of which the general form can be written as

$$
\begin{aligned}
P^{(n)}(t) = \int_0^\infty \mathrm{d}t_n \int_0^\infty \mathrm{d}t_{n-1} \cdots \int_0^\infty \mathrm{d}t_1 R^{(n)}(t_n, t_{n-1}, \ldots, t_1) \\
\times E(\mathbf{r}, t - t_n) E(\mathbf{r}, t - t_n - t_{n-1}) \cdots \\
\times E(\mathbf{r}, t - t_n - t_{n-1} \cdots - t_1),
\end{aligned}
\tag{1.1}
$$

where $P^{(n)}(t)$ is the nth order time-dependent nonlinear response function of the medium. The fifth-order polarization is thus given by

$$
\begin{aligned}
P^{(5)}(\mathbf{r}, t) = \int_0^\infty \mathrm{d}\tau_4 \int_0^\infty \mathrm{d}\tau_2 E(\mathbf{r}, t) |E(\mathbf{r}, t - \tau_2)|^2 \\
\times |E(\mathbf{r}, t - \tau_2 - \tau_4)|^2 R^{(5)}(\tau_2, \tau_4),
\end{aligned}
\tag{1.2}
$$

where the pump and probe pulses are represented by their electric fields, $E(\mathbf{r}, t)$, occurring at time delays τ_2 and τ_4. The polarization is dependent on the fifth-order response of the sample:

$$R^{(5)}(\tau_2, \tau_4) = \left(\frac{i}{\hbar}\right)^2 \langle [[\alpha(\tau_2 + \tau_4), \alpha(\tau_2)], \alpha(0)]\rho_{eq}\rangle, \qquad (1.3)$$

where square brackets denote commutators, the angled brackets denote an ensemble average, $\alpha(\tau)$ is the polarizability operator at time τ, and ρ_{eq} is the equilibrium density matrix. The response is a two-time interval correlation function. If the laser pulses are assumed to be delta functions and the slowly varying envelope approximation is made then the polarization is proportional to the response. The polarization will radiate an electric field which is the resulting measured spectroscopic signal:

$$E_s(t) = 2\pi i \frac{\omega_s^2}{k_s c^2} \int_0^L dz P^{(5)}(z, t) e^{i\Delta kz}. \qquad (1.4)$$

Here ω_s is the signal field radial frequency, k_s is the signal wave vector, c is the speed of light, L is the sample pathlength, and Δk is the wave vector mismatch for generation of the signal. This mismatch arises due to the difference between the wavevector, k_s, and the wavevector in the medium:

$$k'_s = \frac{\omega_s}{c} n_s, \qquad (1.5)$$

where n_s is the index of refraction of the medium. Introducing the response function and electric fields in place of the polarization results in

$$E_s(\tau_2, \tau_4) \propto i \frac{\omega_s}{n_s} \int_0^L dz E_1 E_2 E_3 E_4 E_5 R^{(5)}(\tau_2, \tau_4) e^{i\Delta kz}, \qquad (1.6)$$

Upon integration along the propagation direction z from sample pathlength 0 to L the fifth-order Raman signal becomes

$$E_s(\tau_2, \tau_4) = i \frac{L\omega_s}{n_s} E^5 R^{(5)}(\tau_2, \tau_4) \frac{\sin(\Delta kL/2)}{\Delta kL/2} e^{(i\Delta kL/2)}, \qquad (1.7)$$

where for convenience the electric fields of the five laser pulses have been assumed to be identical. One of the primary assumptions made here was that all of the laser beams generating the electric fields are propagating collinearly. This is only valid if the angles between the incoming beams are very small, essentially making them parallel as they move through the sample. From the above equation we can make note of several factors that will influence the signal strength of the fifth-order Raman spectroscopy. The first is that the signal grows linearly in proportion to the sample pathlength and the signal frequency. The second is that the smaller the Δk value is, the better phase-matched the nonlinear process is. As would be expected there is also a strong dependence on the electric field of the laser used. All of these factors influence the signal strength, and all of these factors are experimentally adjustable making them critically important to keep in mind while designing a fifth-order Raman experiment.

1.2.1 Origin of the Fifth-Order Raman Signal

The fifth-order Raman signal can be generated through a number of different interactions, several of which were not immediately obvious in the theory's infancy. The original understanding was that the signal could be generated by a nonlinear dependence of the nonlinear polarizability on nuclear coordinate

$$\alpha(q) = \alpha(q_0) + \sum_i \left(\frac{\partial \alpha}{\partial q_i} \right)_{q_0} + \frac{1}{2} \sum_{ij} \left(\frac{\partial^2 \alpha}{\partial q_i \partial q_j} \right)_{q_0} q_i q_j + \cdots . \tag{1.8}$$

The first term in this Taylor expansion around q_0 has no coordinate dependence and is time independent. The second term is linear in coordinate dependence and corresponds to the signal measured in third-order Raman experiments. To generate a fifth-order Raman signal the third term is necessary: if a single harmonic mode, q_j, is assumed, signal generation requires a two-quantum coherence to be created by one of the interactions. Subsequent to this realization it was proposed [28] that the spectroscopy also probes any anharmonicity in the vibrational potential, which corresponds to the inclusion of a cubic anharmonicity term:

$$V(q) = \frac{1}{2} \sum_i \beta_i^{(2)} q_i^2 + \frac{1}{6} \sum_{i,j,k} \beta_{ijk}^{(3)} q_i q_j q_k + \cdots . \tag{1.9}$$

$\beta_i^{(2)}$ are the harmonic force constants and $\beta_{ijk}^{(3)}$ are the strengths of the cubic term. This result stems from the notion that the ground-state Hamiltonian is not time-independent and is in fact evolving in time. The harmonic modes can thus change energy as the potential surface evolves, sometimes called dynamical anharmonicity. Each of these two contributions to the fifth-order Raman signal can be separated, allowing the possibility for modeling of the experiment to accurately evaluate the real signal's dependence on each source term [29].

The most recent contribution to the signal evaluation has been the observation that in a harmonic set of modes mode coupling can occur through the polarizability. In the case as depicted in (1.8) the two modes q_i and q_j can couple through the nonlinear polarizability term. This is the only source of mode coupling between the orthogonal harmonic modes and has been termed *mode mixing through polarizability*. This has led to some interesting proposals, including the idea that it should be possible to observe mode coupling between intramolecular Raman active modes.

1.2.2   Theoretical Results Summary

A large amount of effort has been devoted to evaluating the fifth-order Raman signal for a variety of systems. The first method used in such simulations is the *multi-mode Brownian oscillator* (MMBO) model [16] which approached the problem as a system of oscillators coupled to a heat bath. The oscillators are obtained from a fit of the low-frequency spectrum of the liquid in question:

$$R^{(3)}(\nu) = \sum_i \frac{\nu A_i C_i}{\left(B_i{}^2 - \nu^2\right)^2 + \nu^2 C_i{}^2}, \tag{1.10}$$

where A, B, and C are fit parameters and ν is the frequency in cm^{-1}. In the case of water [30] the fit was made to the imaginary component of the Fourier transform of the deconvoluted femtosecond optically heterodyne detected, OKE experiment. The best fit required six oscillators, of varying magnitudes and frequencies. These oscillators were then inserted into the calculations as either the inhomogeneous distribution of oscillators or the homogeneous linewidth, depending on which limiting case was being examined. This simple model predicts the signal characteristics for the two extreme cases: (1) the inhomogeneous case has a strong response along the time diagonal, corresponding to an echo-like rephasing of the oscillators and (2) a distinct lack of an echo signal in the homogeneous case, in fact little signal along both the pump delay axis (τ_1 in Fig. 1.4) and the diagonal but a noticeable signal along the probe delay axis (τ_2 in Fig. 1.4). Since real systems would likely be an intermediate situation between the two limiting cases it was not immediately evident how useful this modeling would prove to be though it was clear that if a liquid's low frequency modes were strongly inhomogeneously broadened, it would be obvious in the fifth-order signal.

Once the anharmonicity of the potential as a possible source for the fifth-order Raman signal was taken into account, some modifications were made to the MMBO model. In the simplest case the anharmonicity was introduced as a perturbation to the harmonic vibrational modes (see (1.9)) [31]. This model proved to have the same convenience in interpretation of the limiting cases as the simple MMBO model but with the added possibility of distinction between signal generated by the nonlinearity in the polarizability as opposed to signal generated by the anharmonicity.

Perhaps the most comprehensive early theory work was performed by Tanimura and collaborators [31–33]. They performed numerous calculations using a variety of potential models, heat bath models, and temperatures. In each case the results were thoroughly examined to determine the signal contributions from the various possible sources. Tanimura's most important conclusions were that no matter the system studied or the oscillator models used, the signals tended to have clear signatures that should allow comparison

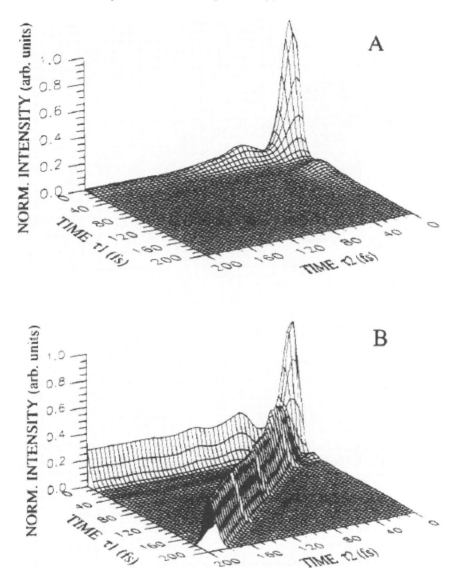

Fig. 1.4 Fifth-order Raman signal intensity calculation for water based on the MMBO method [30]. (**a**) homogeneous limit, (**b**) inhomogeneous limit. τ_1 is the pump delay, τ_2 is the probe delay. Reprinted with permission from [30]. Copyright 1994, American Chemical Society

to experiment and real systems. An example of this is shown in Fig. 1.5 where a strongly damped Morse oscillator coupled to a Gaussian-white noise bath was used as the potential model [33]. The sole source of the 2D Raman signal in Fig. 1.5 is the anharmonicity in the potential since the polarizability coordinate dependence has been limited to the linear term. The result is a signal that is

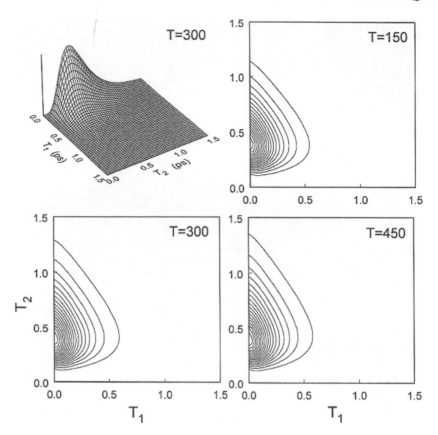

Fig. 1.5   2D Raman signal intensity of a Morse potential with linear polarizability in the strong damping case; T is temperature in Kelvin [33]. Reprinted from [33]. Copyright 1998, with permission from Elsevier

elongated along the probe delay (T_2 in Fig. 1.5), while decaying quickly along the pump delay (T_1 in Fig. 1.5). An interesting added feature is the temperature dependence which has a profound impact on the signal along the probe delay. When the fifth-order Raman signal source is changed from the anharmonicity to the nonlinear polarizability by making the potential harmonic and adding the quadratic term to the polarizability, the signal becomes temperature independent in both the strongly damped (see Fig. 1.6) and weakly damped cases. So another predictive tool is provided to the experimentalist: By varying the temperature and observing the response, it may be possible to determine the relative contributions from the anharmonicity compared to the nonlinear polarizability.

The next major contributions to the progress of theory used normal mode [34–36] calculations for various simple liquids. Normal mode theory was

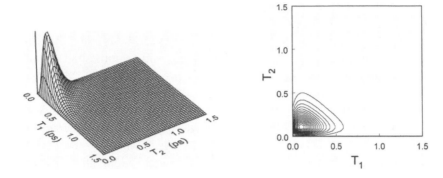

Fig. 1.6 2D Raman signal intensity of a harmonic potential with nonlinear polarizability in the strong damping case [33]. Reprinted from [33]. Copyright 1998, with permission from Elsevier

initially chosen because of its ability to treat the two-time Poisson brackets in the fifth-order response (1.3) and because of its ability to relate signal components to physical phenomena. The initial simulations of Saito and Ohmine used both *quenched normal mode* (QNM) and *instantaneous normal mode* (INM) theories to show that the mode mixing through polarizability has the potential to reduce the echo signature of inhomogeneous broadening [37]. A caveat to this conclusion is that their simulations for both water and liquid carbon disulfide (CS_2) used a harmonic approximation which ignored the anharmonic contributions to both the mode mixing and the fifth-order signal. Interestingly, though the mode mixing had a definite impact on the echo-signal amplitude it did not eliminate it completely and their carbon disulfide response had a clear signal along the time diagonal for both the QNM and INM models. Concurrent to this work, Murry, Fourkas, and Keyes performed an analysis of the third- and fifth-order response from a polarization perspective [38]. Their INM calculations generated *polarization-weighted density of states* (PWDOS) for various polarization tensor elements of CS_2. Essentially by varying the polarizations of the six fields involved in the experiment (four pump pulses, one probe pulse and the signal) it would be possible to analyze different contributions to the response. Based on the results of these calculations they concluded that there is no relationship between the third-order PWDOS and the fifth-order PWDOS and that in order to properly analyze the fifth-order response a full characterization of all the polarization tensor elements would be necessary. Fourkas and Keyes [39] then developed an INM theory with the harmonic approximation replaced by an ensemble average and replaced the coordinate variable, q_α, with the force variable, F_α. This allowed them to provide a theoretical framework for both third and fifth-order calculations. Their third-order calculations showed decent experimental agreement at short times (up to $\sim$300 fs) and more accurate agreement at longer times than previous theories had managed, however no fifth-order calculations were performed. The drawbacks to their approach were that no rotational–diffusional motions were included, decreasing accuracy at long times ($>500\,\mathrm{fs}$), and that they had not included the mode mixing terms that Saito and Ohmine had found to be critically important. Ma and Stratt entered the arena in 2000 by publishing a paper [40] that compared the third and fifth-order results obtained using molecular dynamics (MD) simulations with those obtained using INM. To simplify the MD calculations liquid xenon was chosen. The results are striking; the MD simulation [41] shows no rephasing signal whatsoever (see Fig. 1.8) while the INM calculation [42] has a strong signal along the diagonal (see Fig. 1.7). Clearly the two methods indicate almost diametrically opposite results in terms of the intermolecular potential. Liquid Xe is an atomic liquid that intuitively should have weak intermolecular forces when compared with either a molecular dipole interaction or a hydrogen-bonded system yet its INM calculation shows a clear signature of inhomogeneous broadening. Ma and Stratt concluded that the fifth-order Raman signal is extraordinarily sensitive to the dynamical anharmonicity (1.9) which is not accounted for in INM theory. The echo-like signal is a result of the INM calculation's basis on independent harmonic modes and simply reflects the underlying assumptions of the method, not anything intrinsic to the liquid.

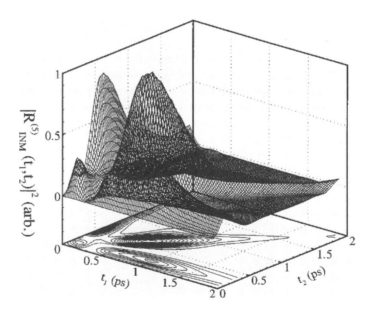

Fig. 1.7 Fifth-order Raman signal intensity for liquid Xe calculated using INM. Note the distinct echo signal along the time diagonal ($t_1 = t_2$) [42]. Reused with permission from [42]. Copyright 2002, American Institute of Physics

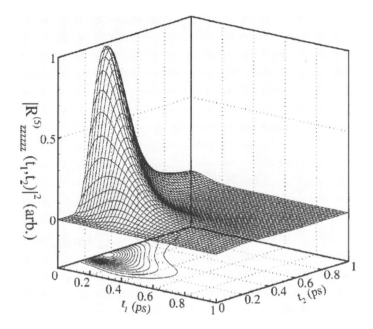

Fig. 1.8 Fifth-order Raman signal intensity for a 32-atom MD simulation of liquid Xe. Note the distinct lack of echo signal along the time diagonal [41]. Reused with permission from [41]. Copyright 2002, American Institute of Physics

This result caused a flurry of interest in the theoretical community with several groups trying to either adapt the INM theory to more accurately predict the fifth-order Raman response or to generate a more accurate theoretical model. Denny and Reichman [43] used a molecular hydrodynamics approach to project the dynamically relevant portions of the polarizability operator onto bilinear pairs of fluctuating density operators. This allowed them to interpret the fifth-order response in terms of density fluctuations. The result of their simulations for liquid Xe successfully removed the diagonal rephasing signal from the INM result (see Fig. 1.9), though it still lacks several characteristics of the MD simulation (primarily the elongated signal along the probe delay, t_2 in Figs. 1.8 and 1.9). They attribute the lack of echo formation to the fact that the density fluctuations in liquid xenon are heavily overdamped, thus accurately matching the MD result. However the lack of signal along the probe delay is troublesome and they attribute this failure to the Gaussian nature of the density modes in the theory and posit that perhaps non-Gaussian corrections must be made to properly account for the dynamics along the probe delay [44]. Since the nature of the calculation generates results that can be interpreted in terms of density fluctuations, one of the suggested avenues of experimental research is that of supercooled liquids where they predict interesting dynamics. A subsequent

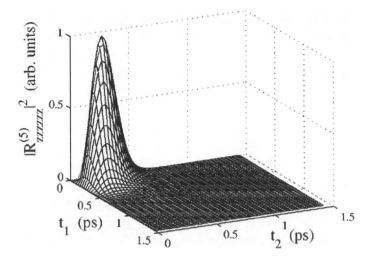

Fig. 1.9   Fifth-order Raman response intensity calculated using the molecular hydrodynamics mode-coupling approach [43]. Reprinted with permission from [43]. Copyright 2001, American Physical Society

effort by Cao, Yang, and Wu [45] also used hydrodynamic theory with a slightly modified Gaussian factorization scheme to calculate the fifth-order response of liquid Xe, achieving similar results to Denny and Reichman. Their results compare favorably with the MD simulation of Ma and Stratt.

Yet another theoretical approach was taken by van Zon and Schofield [46], as well as by Kim and Keyes [47], in which the generalized Langevin equation was used to investigate the problem. Van Zon and Schofield's work uses a mode-coupling formalism and emphasizes the distinction between Gaussian and non-Gaussian noise in the Langevin equation and its effect on multiple-time correlation functions (of which the fifth-order Raman response is an example). Kim and Keyes also attacked the problem using the generalized Langevin equation and they calculated the fifth-order response for liquid CS_2. Their response is difficult to interpret but the salient features are a diagonal peak at just under 200 fs and signal along the probe delay as well as along the pump delay with the restriction that the signal goes to zero along the pump delay axis.

The failure of INM theory is a strong indication of the sensitivity of the fifth-order Raman signal to anharmonicities. Ma and Stratt modified INM theory to allow for dephasing of the modes which they termed *adiabatic instantaneous normal mode theory* [42]. This allowed them to incorporate the dynamical anharmonicity which was missing from their previous calculation (see (1.9)). In an approach similar to Okumura and Tanimura, the response was broken into its component parts to investigate their dependence on the various signal contributions (see Fig. 1.10). When treated individually the dynamical anharmonicity

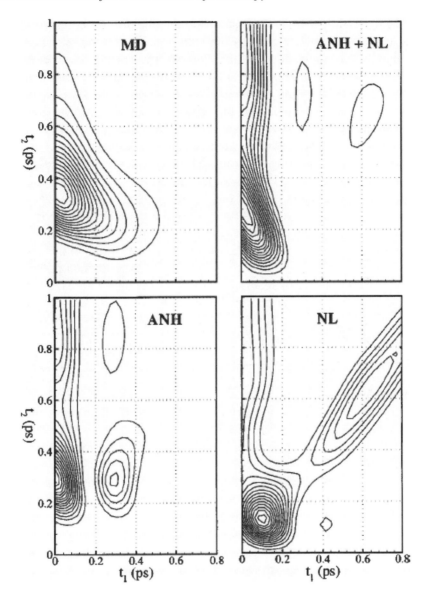

Fig. 1.10 Fifth-order response for liquid Xe calculated using Ma and Stratt's adiabatic INM approach. Note the similarity between the MD simulation (Fig. 1.8) and the combined dynamical anharmonicity (ANH) + nonlinear polarizability (NL) calculation [42]. Reused with permission from [42]. Copyright 2002, American Institute of Physics

results in a ridge along the probe delay t_2 (ANH in Fig. 1.10), whereas when it is ignored the response has a strong diagonal ridge combined with the signal along the probe delay (NL in Fig. 1.10). The resulting full response for liquid xenon (ANH + NL in Fig. 1.10) very closely matches the MD simulation (see Fig. 1.8), leading them to the conclusion that the fifth-order response may prove to be governed much more by the anharmonicity of the intermolecular modes than by the nonlinearity in the polarizability. There are two caveats to this result: (1) Note that the adiabatic INM theory does not include the mode coupling through polarizability that Saito and Ohmine predicted would suppress the echo signal. (2) The signal along the probe axis contains information about the population relaxation of the excited modes, the adiabatic INM calculation does not allow for exchange of energy between modes leading to the conclusion that the long-lived signal along the probe delay may not accurately describe the liquid's dynamical response. Their most recent results [48] have continued this approach of using the INM calculations in comparison to MD to analyze the intermolecular dynamics. Ma and Stratt inserted a single, rigid CS_2 molecule into a bath of 29 Xe atoms for their MD simulations and separated the INM influence spectra into five rotationally invariant contributions to the response functions. By relating these influence spectra to the experimental polarization settings they were able to isolate, in specific cases, the contributions each invariant had on either the nonlinear-coupling or the anharmonic-dynamics portion of the signal. This analysis allowed them to suggest that the $R^{(5)}_{1111mm}$ tensor element only contains contributions from the anharmonic dynamics of the liquid and the $R^{(5)}_{mm1111}$ response looks preferentially at the nonlinear dependence of the polarizability on coordinate, though without the same level of discrimination. They emphasize, however, that the results from the INM calculations for these two tensor elements do not perfectly match the MD simulation and to be cautious about relying on the INM predictions for the total fifth-order response. As has been noted above, several contributions are missing from the adiabatic INM calculation.

While these innovations were being introduced to normal-mode theory, advances were also being made in molecular dynamics by Jansen, Snijders, and Duppen. The approach taken was to simulate the experiment directly: apply an electric field to a finite volume element full of molecules. In the nonresonant case the electric field interacts with the molecules through the polarizability operator, applying a torque to the molecules. This approach is called the *finite-field nonequilibrium molecular dynamics method* (FFMD) [49]. By controlling the number of electric fields and their timing it becomes straightforward to simulate either the third or fifth-order response for any liquid with this approach. The initial results for this type of calculation produced excellent agreement with experiment for the third-order response of CS_2 at long times (the diffusive tail) but poor agreement for short times (<500 fs including the magnitude of the nuclear signal at 200 fs). The conclusion was that the technique was promising and if local-field effects were incorporated, the response would be more accurate at early times. Their first attempt at incorporating many-body

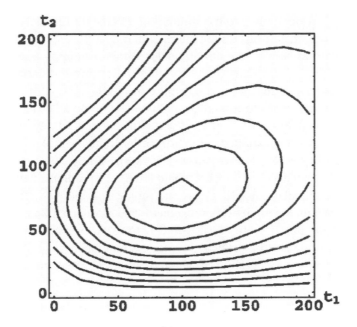

Fig. 1.11 Fifth-order Raman response $R^{(b)}_{111111}$ of liquid CS_2 simulated using the finite-field MD approach. t_1 is the pump delay, t_2 the probe delay, times are in femtosecond [50]. Reused with permission from [50]. Copyright 2001, American Institute of Physics

effects into the FFMD simulation used a *dipole–induced dipole* (DID) model for the local-field interaction. The individual molecules feel not only the effect of the macroscopic electric field of the laser but also the effect of the electric fields generated by the induced dipole moments in the surrounding molecules. Once this was incorporated into the simulation the short-time dynamics of the third-order response for CS_2 was successfully reproduced. Some limited fifth-order simulations for CS_2 were also performed using this model on a box of 256 molecules, the results of which seemed to indicate a slightly off-diagonal peak somewhere around 100 fs ($t_1 = t_2 \sim 100$ fs in Fig. 1.11) [50]. Perhaps of most interest to experimentalists, this was the first fifth-order simulation that explicitly took into account the polarization of the electric field and calculated the response for various different polarization combinations. The next modification to the FFMD simulation was to introduce a more accurate model for the many-body effects. The *direct reaction field* (DRF) model slightly modifies the DID model to also take into account electron wavefunction overlap effects. This more accurate model was initially used to calculate the third-order Raman response for liquid Xe and it proved to be slightly more accurate than the DID model in matching the experimental signal [51]. The real test of the model came when it was applied to both the third-order [52] and the fifth-order Raman response of CS_2 [53].

Four models were applied to the simulation: (1) the first model simply uses the molecular polarizabilities for the interaction induced effects (MOL), (2) the second approximation uses the DID model, (3) the third approximation takes into the account the extended structure of the molecules and uses the atomic polarizabilities to account for induced multipoles in the interaction (POL), and (4) the most complicated and computationally intensive model is the DRF model where electron wavefunction overlap is added to the many-body effects. Each of these models was then applied to the fifth-order Raman simulation of 64 CS_2 molecules (see Fig. 1.12). The response varies surprisingly between the various models, which is not necessarily what would have been predicted from the third-order response calculations. Obviously the interaction-induced effects have a profound impact on the higher order response. The most accurate model (DRF) has several pronounced features. The first feature is the peak that is shifted off the diagonal toward the probe delay; this is similar in general description to the adiabatic INM calculation of Ma and Stratt. The second feature is the

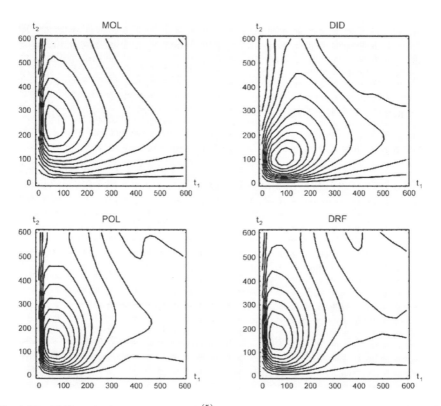

Fig. 1.12   Fifth-order Raman response $R^{(5)}_{111111}$ of liquid CS_2 simulated using the finite-field MD approach with four different interaction-induced effect models (see text for details). The time scale is in femtosecond [53]. Reprinted with permission from [53]. Copyright 2003, American Physical Society

extended signal along the probe delay (>600 fs), while the signal truncates very quickly along the pump delay (<200 fs), causing a pronounced asymmetry in the response. Jansen et al. also simulated a number of other polarization combinations to better supply experimentalists with comparative data. Unfortunately the molecular dynamics simulations do not lend themselves well to physical interpretation, it is not clear what the source for the various signals is, and little insight can be gained without further analysis.

While the FFMD simulations were progressing Saito and Ohmine were working on a full MD simulation [54]. Their simulation was performed using flexible molecular potentials and 32 molecules for both CS_2 and water, the DID approximation was used for the total polarizability of the system (Fig. 1.13). The results of this work were intriguing. The fifth-order response of CS_2 is similar in overall shape to that of Jansen et al. with one discrepancy: The signal changes sign along the probe delay, creating a node in the signal at around 200 fs. This is an unusual finding as it implies an interference effect between different dynamical signal contributions that must nearly cancel to give such a node so displaced from the $t = 0$ origin. Saito and Ohmine attribute it to intermolecular dynamics beyond normal mode dynamics, specifically the couplings between rotational modes in liquid CS_2. Basically, the anharmonic terms in the internuclear potential are opposite in sign to the nonlinear electronic polarizability. In contrast to the work of Ma and Stratt, it would seem from this work that these two contributions are of similar magnitude and of opposite sign. Saito and Ohmine also calculated the *constant temperature velocity reassignment echo* (CVRE) which allows them

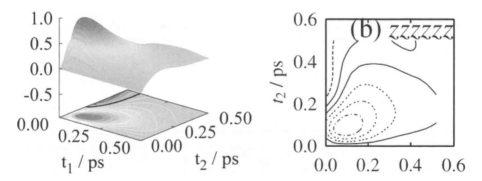

Fig. 1.13 Fifth-order Raman response $R^{(5)}_{111111}$ of liquid CS_2 simulated using molecular dynamics. There is a nodal line along the probe delay (t_2) as indicated by the *solid line* [54, 55]. *Bottom figure* reprinted with permission from [54]. Copyright 2003 by the American Physical Society. *Top figure* reused with permission from [55]. Copyright 2003, American Institute of Physics

to assess the contributions to the signal from the dynamical anharmonicity. By comparing the CVRE signal and the fifth-order response of CS_2 they determined that the echo in the response is being suppressed by the mode coupling through polarizability. The fifth-order response of water is also presented but with little analysis except a comparison to their previous normal-mode calculation. A more in depth study followed where the fifth-order response was calculated for a variety of polarization conditions for CS_2 [55]. In addition the DRF model was also used for comparison to the DID model used previously. The overall dynamics of the DRF simulation for CS_2 remains remarkably unchanged from the DID simulation, in contrast to the results obtained by Jansen et al. where distinct differences were seen depending on which model was used.

The most recent contribution to the field of nonlinear response theory comes from Keyes, Space, and collaborators [56, 57]. Their approach results in an exact classical response function written in terms of classical time correlation functions (TCF). The response takes into account the nonlinear polarizability and is used in a fully anharmonic MD simulation to simulate the fifth-order response of CS_2 (see Fig. 1.14). The results are strikingly similar to those obtained by Jansen et al. as well as to the simulations, both MD and adiabatic INM, of liquid Xe (see Figs. 1.8 and 1.10). The dominant features are the ridge along the probe delay and the distinct lack of signal along the pump delay.

The theoretical community has fully embraced the fifth-order Raman experiment and its unmatched potential for testing theoretical models for low-frequency motions in liquids. Since the advent of the spectroscopy, a substantial body of calculations and simulations has been generated for a variety of systems. In several cases theories have had to be completely revised to attack the problem. With this plethora of theoretical data available for comparison the focus now must shift to the experiment since in several cases the experiment is what has driven some of the theoretical decisions.

1.3   Experimental Advances

The fifth-order Raman experiment has proven to be a challenging experimental endeavor. Beyond the difficulties inherent in performing an ultrafast nonlinear spectroscopic experiment requiring four pump pulses and one probe pulse to be carefully spatially aligned into a thin liquid sample, the presence of unwanted parasitic signals has proven to be a substantial problem [18–20]. The initial fifth-order Raman experiments performed proved to be measurements of these unwanted signals with little discernible fifth-order response content. These so-called cascaded signals are now well documented and are a result of two types of third-order Raman signals, respectively, called *sequential cascade* and *parallel*

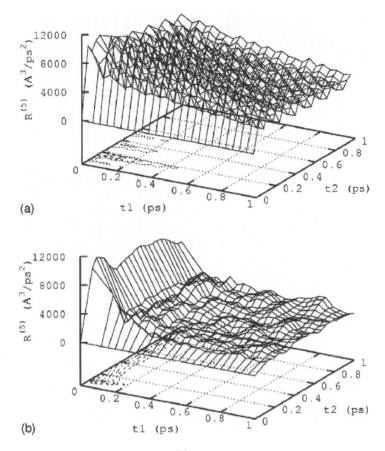

Fig. 1.14 Fifth-order Raman response $|R^{(5)}_{zzzzzz}|$ of liquid CS_2 calculated using a time correlation function method formulated by Keyes et al. with two different TCF time steps: (*a*) 0.008 ps and (b) 0.04 ps. [56]. Reused with permission from [56] Copyright 2003, American Institute of Physics

cascade (see Fig. 1.15). The sequential cascade is a result of a third-order event occurring on one chromophore, with the signal photon mixing as a pump photon in a second third-order event on another chromophore. The parallel cascade involves a third-order event generating a signal photon, which then acts as the probe photon on another chromophore. The result from either cascaded process is a signal being generated in the same geometric phase-matched direction and with the same power dependence as the fifth-order signal.

Historically, the first experiments that helped stimulate the theory were essentially multipulse OKE echo experiments without specific phase matching [15]. The first experiments deliberately exploiting phase matching to isolate the fifth-order signal were performed by Tominaga and Yoshihara [58]. These initial experiments were performed on liquid CS_2 at room temperature. This was

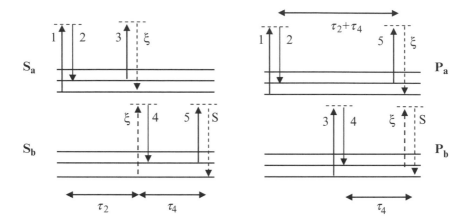

Fig. 1.15 The energy level diagrams corresponding to the two different cascaded signal contributions in the fifth-order Raman experiment. S_a is the first sequential cascade step, S_b the second sequential cascade step, P_a the first parallel cascade step, and P_b the second parallel cascade step. Each step is on a different chromophore

quickly followed by several other groups who all achieved consistent results for liquid CS_2, with several attempts made on other liquids such as benzene [59], $CHCl_3$, and CCl_4 [60] (the latter two involved intramolecular mode excitation). Within 6 years of the birth of the spectroscopy the number of experiments having been performed was impressive [61–68]. In each case, many different methods of analysis were used and comparison was made to the existing theory work of the time. Several discrepancies were noted between the inability of theory to predict the experimental results and an analysis made by Steffen and Duppen foreshadowed the problems that lay ahead [69]. It was then discovered that the signal being measured was not the pure fifth-order response. In fact an experimental analysis performed by one of the groups [20] indicated that in every case the signal measured was a previously unaccounted for third-order cascaded signal which they named the parallel cascade. Frustratingly, the groups had all been aware of the possibility of these lower order parasitic signals and had carefully designed their beam geometries to avoid a similar parasitic signal, the sequential cascade, but had neglected the parallel cascade channel. It was realized at that point that the rephasing process involved a two-quantum transition and that the signal would be extremely small relative to conventional third-order signals that our group started to develop the diffractive optics for nonlinear spectroscopy with the primary motivation to directly observe the fifth-order Raman response. The demands of the experiment for the highest sensitivity possible, stringent phase matching geometries to discriminate against much larger parasitic lower order signals, and phase-locked signal detection make the fifth-order Raman experiment arguably the most challenging experiment in nonlinear spectroscopy. However, as can be appreciated from the above discussion of the great theoretical interest and exquisite sensitivity of the observable to the liquid state, the advancements in nonlinear spectroscopy required to meet this challenge are worth the effort.

1.3.1 Laser Source and Experiment Synopsis

The laser source used in these experiments is a standard Titanium:Sapphire, regeneratively amplified femtosecond laser that produces 800 nm, 400 μJ, 70 fs pulses at a 1 kHz repetition rate. A small fraction of this output is picked off and used as the 10 μJ, 800 nm probe beam while for the two-color experiment the rest of the light is put through a 300 μm thick BBO doubling crystal, producing 70 μJ, 400 nm pulses which will be used as the pump pairs. Both the pump and probe beams are put through a pair of pre-compensation prisms to ensure short pulses reach the sample [70]. The probe beam is then sequentially put through (1) a retroreflector mounted on a motorized, computer-controlled delay stage to control the probe time delay τ_4, (2) a waveplate/polarizer combination for intensity control, (3) a waveplate to control the orientation of the probe polarization, and (4) a lens to focus the beam into a 250 μm diameter spot on the diffractive optic which will generate the beam pattern. The pump beam is split into two equal beams using a beamsplitter, producing the two pump arms. One pump beam is bounced off a retroreflector mounted on a motorized, computer-controlled translation stage to control the pump time delay τ_2. The second beam is bounced off a retroreflector mounted on a manual translation stage which is used for coarse time-overlap adjustment. Each 400 nm beam is then put through its own version of the same sequence of optics as the probe beam. The result is three almost collinear beams, two blue and one IR, focussed tightly onto the back surface of a transmission diffractive optic. A second experimental setup which has also been used takes the undoubled leakage from the BBO crystal and uses it to pump a homebuilt noncollinear optical parametric amplifier [71] (NOPA) which is then compressed using chirped mirrors. This leads to a three-color experiment where the two pump pairs are now 530 and 400 nm, respectively, while the probe pulse is 800 nm. This latter configuration represents the most general approach, allowing multicolor discrimination against background scatter, and is shown in Fig. 1.16.

The custom-made diffractive optic generates three exact replicas (two blue, one IR in the two-color experiment) of the required phase-matching beam geometry with slight horizontal separation between blue patterns and a large angular diffraction separating out the IR pattern. Using filters and irises, the proper beam geometry for the phase-matching condition can be chosen. This results in two time-coincident blue beams generated by the manual translation stage arm, two time-coincident blue beams generated by the τ_2 motorized translation stage arm, and two time-coincident IR beams generated by the τ_4 motorized

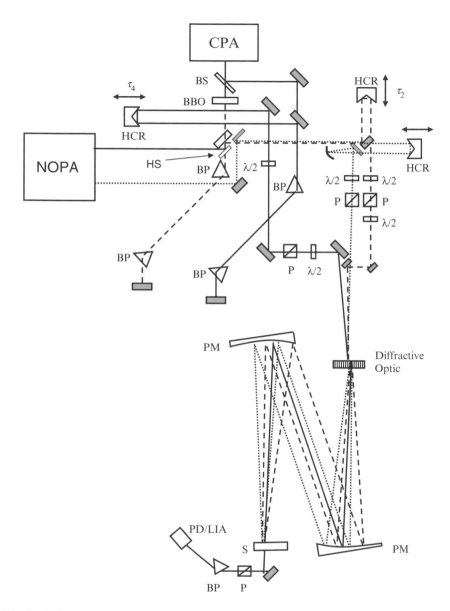

Fig. 1.16 The most general three-color experiment. HS harmonic separator, BS beam splitter, BBO β-barium borate crystal, HCR hollow cube retroreflector, BP Brewster prism, CS cover slip, $\lambda/2$ half wave plate, P polarizer, PM parabolic mirror, S sample, PD/LIA photodiode/lock-in amplifier. The *solid lines* indicate 800 nm beams, the *dotted lines* are the 530 nm beams, and the *dashed lines* are the 400 nm beams. The intervening optics between the DOE and the sample are not shown for clarity. Half-wave plates and phase plates are introduced in the straight section between the parabolic mirrors to enable the selection of any $\chi^{(5)}$ tensor element, mixed polarization state, or to perform a phase scan

translation stage arm. One of the IR beams is generated by the diffractive optic so it overlaps perfectly with the anticipated signal direction, which greatly facilitates alignment and allows for the possibility of heterodyne detecting the signal (note that the signal wavelength will be the same as the probe: 800 nm). The six beams are then reflected off two off-axis parabolic mirrors onto the sample cell (see Fig. 1.16). This has the effect of re-creating the grating phase profile in the sample, ensuring that all phase-matching conditions are met. When either of the pump beams is time overlapped with the probe beam at the diffractive optic OKE signal can be generated in the glass of the diffractive optic. Since this is not a desirable effect, the pulses are kept temporally separated at the optic. The pulses need to be time overlapped at the sample and so glass is placed into the beam path to delay the pulse and fine control is obtained through tilting thin cover slips to control the amount of glass the beam goes through. This allows the five beams to be time coincident at the sample. The signal/reference beam is then put through an aperture followed by a polarizer to control the signal polarization. Finally a prism is used to separate off any scattered blue light and the beam is then focussed onto a linear photodiode. One of the pump beams is chopped and a lock-in amplifier synchronized to the chopper is used to acquire the signal and input it into a computer. A scan is performed by acquiring signal at each position of either of the two motorized translation stages. In general a raster scan was performed where a τ_2 trace was taken; the τ_4 stage was moved some increment (usually 25 or 50 fs), and τ_2 was rescanned until the two-dimensional time region was completed. A diagonal scan ($\tau_2 = \tau_4$) can be performed by scanning both translation stages simultaneously. The samples used are spectroscopic-grade CS_2 and benzene that are pumped through a flow cell with an optical pathlength of 500, 300, or 100 μm using a peristaltic pump. The pathlength is controlled by Teflon spacers between the two compressed cell windows. The front window of the cell is a 100 μm thick microscope cover slip, kept thin so as to keep the group-velocity dispersion between the multicolor pump and probe pulses to a minimum, and the back window is a 1 mm thick microscope slide. Since the index of refraction of the various optical pathlength components (glass and liquid sample) are different for the various laser wavelengths used, the pump and probe pulses have different velocities in the cell. This can result in a time smearing that reduces the temporal resolution of the signal. To avoid this sample, path lengths are kept as short as possible while still maintaining measurable signal.

1.3.2 Diffractive Optics

One of our major contributions to the experimental techniques used in nonlinear spectroscopy has been the wide-ranging use of diffractive optics. Simple period gratings have been used in third-order resonant [22, 72, 73] and nonresonant

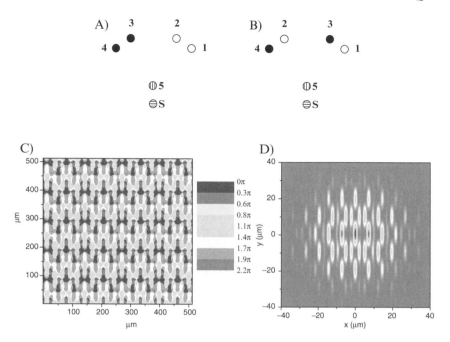

Fig. 1.17 Two fifth-order phase-matching geometries formed from the same diffractive optic pattern designed to specifically enhance contrast against lower order cascades and determine the relative magnitudes of such cascades. Pulses with the same pattern are time coincident. (a) baseball geometry, (b) crossed-beam geometry, (c) 500 μm × 500 μm section of the phase profile of the actual diffractive optic, (d) effect of using too small a probe beam to sample the entire fifth-order polarization in the sample. In the case shown, diffraction would only involve approximately three fringes in the fifth-order pattern, yielding poor diffraction, whereas the simple-period third-order polarization components would have many fringes. The result is that the inherent contrast of the phase matching geometry would be significantly reduced

experiments [21,74] but the fifth-order phase-matching geometry is much more complicated and requires a custom *diffractive optical element* (DOE) [11]. To generate the desired beam pattern for proper phase matching and high contrast against lower order cascades (Fig. 1.17), the DOE we designed contains 20 Fourier components. The amplitude of the different components was adjusted and diffraction efficiency calculated across the entire visible spectrum using coupled wave equations to provide the maximum efficiency with the greatest spectral bandwidth for femtosecond studies. A 500 μm × 500 μm section of the phase profile of the DOE is shown in Fig. 1.17c. One can immediately appreciate the complexity of the optic required to give the designed beam parameters. More importantly, the spatial profile of the diffractive optic is instructive in its own right. This spatial pattern is essentially identical to the polarization that is induced in the material by the various timed laser field interactions. Inspection of this pattern reveals that there are relatively long spatial components that must be sampled by the probe beam to give good diffraction from the five field interactions and truly sample the $\chi^{(5)}$ susceptibility of the material. In conventional four-wave mixing experiments, a general rule of thumb is to design the pump and probe spot sizes such that the probe diffracts from at least 15 fringes for reasonable diffraction efficiency (much like proper filling of a grating). However, if one only uses the beam angles to calculate the dominant Fourier component of the excitation pulse pairs, the fifth-order grating will be underfilled and the contrast against lower order cascades will be lost. The probe spot size needs to be large enough to sample several periods of the complex induced polarization (see Fig. 1.17d for an example of a situation that does not meet this criterion). Part of the reason for the differences between results from different groups reporting $\chi^{(5)}$ signals after the problem with parallel cascades was properly identified is likely due to this effect.

The first advantage the DOE has over other approaches is it makes beam alignment extremely simple, just point your beam at the optic and your beam diffracts into the desired pattern, all that is then required is to focus the beams down into the sample. This avoids much of the alignment issues involving beam splitters and separate alignment of each beam. In fact the number of alignment mirrors necessary to maximize the signal is cut in half. The other major advantage of the DOE-based approach is the added ability to generate a passively phase-locked heterodyning field that perfectly overlaps the signal direction. Designing the DOE to diffract a beam in the signal direction not only makes spatially locating the signal much simpler, but also provides a beam that is phase locked to the probe beam. The phase-locked nature of the beams is guaranteed by the imaging system that has all six beams (four pump, probe, and reference) pass through all the same transfer optics. Any noise in the optical system is imparted to all the beams equally such that the noise is correlated and the phase relationship of all the beams after the diffractive optic is conserved to provide phase-locked pulse pairs. The major independent noise source is due to air currents and this can largely be minimized by the use of enclosures. Phase stability among all six fields, as required for phase-locked signal detection, is better than $\lambda/50$ at 400 nm for periods of hours. This accomplishment would be extremely difficult to do with all active feedback loops as this would require three independent active phase-locked interferometers. The passively phase-locked feature of the diffractive optic approach is a very significant advantage as it increases the signal to noise by several orders of magnitude by allowing measurement of the signal field directly and, equally importantly, enables the use of phase cycling as performed in NMR to separate different signal source terms (vide infra). The main disadvantage of using a DOE is the wasted pulse energy. Our specific DOE has excellent diffraction efficiency of up to 70% across the visible spectrum but since each beam generates a full

beam pattern and only a few beams are used from each diffraction pattern, the
blocked beams are wasted pulse energy. In spite of this drawback the energy
per pulse is well sufficient for the experiment.

1.3.3   Heterodyne Detection

When a signal field is measured using a square-law detector, such as a photodiode, the signal field's intensity is measured. This means any phase information
contained in the field is immediately lost. The way to retrieve this information
is to interfere a reference field with the signal on the detector. This is called
heterodyne detection and it allows full retrieval of the phase and amplitude of
the signal field. In the case of a signal field, $E_{sig}(t)$, mixed with a reference
field, $E_{ref}(t)$, the following expression is obtained

$$
\begin{aligned}
I_{het}(t) &= |E_{sig}(t) + E_{ref}(t)|^2 \\
&= |E_{sig}(t)|^2 + |E_{ref}(t)|^2 + 2E_{sig}(t)E_{ref}(t)\cos\Delta\phi.
\end{aligned}
\tag{1.11}
$$

In this equation $|E_{sig}(t)|^2$ is the signal intensity, or the homodyne signal mea-
sured when the reference field is blocked, and $|E_{ref}(t)|^2$ is the reference field
intensity, which can be either ignored if it has no time dependence or measured
and subtracted. The final term in the equation is one of importance since it
contains the signal field. We can control the phase of the measured signal by
changing the relative phase of the signal and reference fields with respect to
each other. This control over $\Delta\phi$ is obtained by placing a thin glass cover slip
in the reference beam after the diffractive optic. By rotating the angle of the
cover slip we control the relative phase of the reference beam with respect to
the signal beam, allowing us to interfere constructively or destructively with the
signal field. To properly account for all source terms in (1.11) three signals are
measured for each time slice: (1) the signal with both the reference and probe
unblocked (heterodyne signal), (2) the signal with the reference blocked and
probe unblocked (homodyne signal), and (3) the signal with the probe blocked
and reference unblocked (this contribution can result from the interaction of the
single chopped pump beam with the reference). The signals from (2) and (3)
are then subtracted from (1) to obtain the signal field. In addition to the multiple
time scans we also cycle $\Delta\phi$ through a full π phase shift between the signal and
reference fields. This allows us to see the change in phase of the various signal
components, allowing us to identify the various contributions to the signal field.
This is an important feature since it is predicted that the cascaded signal will
be phase shifted from the fifth-order Raman signal [23, 75, 80].

  The ability to heterodyne detect the signal is extremely important in nonlinear spectroscopy where the phase of the signal field often contains information about what physical process generated it. In third-order spectroscopy
heterodyne detection allows the separation of the resonant from the nonresonant

components of the signal. In the fifth-order experiment the signal is explicitly nonresonant and so only one phase component should exist, but as there exists the potential for several contributions to the signal, some desired and some not, each with potentially a different phase, it seems critically important to be able to heterodyne detect the signal. The DOE provides us with this feature free of charge and allows us to passively phase lock the reference field with a stability of $\lambda/50$.

1.3.4 Multicolor Approach

In comparison with other fifth-order Raman experiments we are unique in that we use multiple colors for our pump and probe pulses. The decision behind this was twofold: First, the Raman scattering cross section is much higher for shorter wavelengths, in fact an expected increase of 2^4 is expected [76] for our 400 nm pump pulses when compared to an all 800 nm experiment. The second major advantage to the multicolor approach is the signal field is a completely different wavelength from the pump pulses. This allows spectral filtering to remove the pump pulse scatter which would otherwise make signal detection much more difficult. One possible drawback in using 400 nm light for our pump pulses is this wavelength approaches a two-photon absorption in CS_2. We have carefully characterized our third-order signal, where any two-photon absorption should be visible, and are satisfied this is not a problem in our experiment. As an added check we also performed the experiment using a NOPA to generate the second pair of pump pulses, the results of which are discussed further on.

1.3.5 Cascaded Signals

The simplest way to help isolate the fifth-order signal from these contaminants is to ensure the phase-matching geometry chosen maximizes the fifth-order signal while minimizing the third-order signal. The phase-matching requirement for the fifth-order signal is $k_s = (k_1 - k_2) - (k_3 - k_4) + k_5$, for the intermediate cascade steps the equations are $k_{s1} = k_2 - k_1 + k_3$, $k_{s2} = k_1 - k_2 + k_4$, $k_{p1} = k_1 - k_2 + k_5$, and $k_{p2} = k_4 - k_3 + k_5$. An example of two possible fifth-order beam geometries is given in Fig. 1.17a and b for the particular beam geometries used with the DOE, these representations use the k_x and k_y vectors as the coordinate axes, with the k_z unit vector assumed to be roughly one (if the incoming angles are kept small this is a very good approximation, see Table 1.1).

The two geometries shown have been labeled the *baseball geometry* and the *crossed-beam geometry*. This specific diffraction pattern is that of our custom diffractive optic. Using these beam geometries it is possible to construct the cascaded signals' spatial beam pattern, as shown in Fig. 1.18. From this figure it is clear that what is occurring for each cascaded signal is two separate third-

Table 1.1 Angles in the sample for the four pump beams (1–4) and probe beam (5) in the phase-matching geometry from Fig. 1.17. The beam propagation direction is assumed to be the z-axis, with the xy plane representing the sample front face. θ is the elevation from the z-axis and ϕ is the azimuthal angle measured counterclockwise from the positive x-axis

	Beam 1	Beam 2	Beam 3	Beam 4	Beam 5
θ (deg.)	2.78	1.2	1.2	2.78	3.27
ϕ (deg.)	0	33	147	180	−90

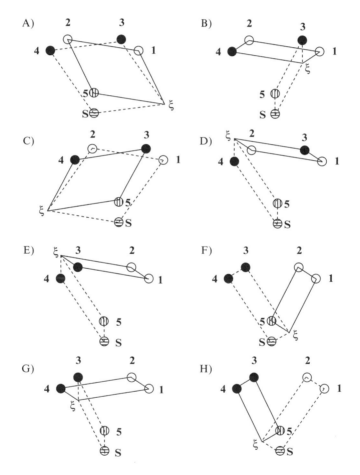

Fig. 1.18 Phase-matching geometries with third-order cascaded signals (ξ) indicated. (**a**) Crossed-beam geometry parallel cascade #1 (**P1**), (**b**) crossed-beam geometry sequential cascade #2 (**S2**), (**c**) Crossed-beam geometry parallel cascade #2 (**P2**), (**d**) Crossed-beam geometry sequential cascade #2 (**S2**), (**e**) Baseball geometry sequential cascade #1 (**S1**), (**f**) Baseball geometry parallel cascade #1 (**P1**), (**g**) Baseball geometry sequential cascade #2 (**S2**), (**h**) Baseball geometry parallel cascade #2 (**P2**). The first step in the cascade (**a**) is represented with *solid lines*, the second step (**b**) with *dashed lines*

order processes being generated using a skewed "box-car" geometry which is often used in transient grating experiments.

Now that the parasitic signals' origin has been established it becomes important to determine what can be done to eliminate them to measure the desired fifth-order Raman signal. The obvious first step is to try to poorly phase match the cascaded signal. This is done starting from the beam geometry angles (Table 1.1) and calculating the unit vectors along each axis using: $\widehat{\mathbf{k}_x} = \sin\theta\cos\phi$, $\widehat{\mathbf{k}_y} = \sin\theta\sin\phi$, and $\widehat{\mathbf{k}_z} = \cos\theta$. The wavevector can then be calculated using

$$\mathbf{k_i} = \widehat{\mathbf{k_i}}\frac{n(\lambda)}{\lambda}2\pi \tag{1.12}$$

where $n(\lambda)$ is the index of refraction at the laser wavelength, λ. The wavevector mismatch can then be calculated from

$$\Delta k = \sqrt{\mathbf{k_x}^2 + \mathbf{k_y}^2 + \mathbf{k_z}^2} - \frac{n(\lambda)}{\lambda}2\pi. \tag{1.13}$$

The results of this calculation are shown in Tables 1.2 and 1.3 for liquid CS_2 and liquid benzene, respectively, for the two-color experiment.

Table 1.2 Wavevector mismatch values for the direct fifth-order process and both steps of the cascaded third-order processes for both phase-matching geometries. For liquid CS_2 in our two-color experimental setup

Process	Δk_a (cm^{-1}) Baseball	Δk_b (cm^{-1}) Baseball	Δk_a (cm^{-1}) Crossed beam	Δk_b (cm^{-1}) Crossed beam
Direct	−15.889			
Sequential 1	580.64	559.32	580.64	559.32
Sequential 2	−511.42	493.20	−511.42	493.20
Parallel 1	225.48	−240.83	1181.86	−1185.92
Parallel 2	225.48	−240.83	1181.86	−1185.92

Table 1.3 Wavevector mismatch values for the direct fifth-order process and both steps of the cascaded third-order processes for both phase-matching geometries. For liquid benzene in our two-color experimental setup

Process	Δk_a (cm^{-1}) Baseball	Δk_b (cm^{-1}) Baseball	Δk_a (cm^{-1}) Crossed-beam	Δk_b (cm^{-1}) Crossed-beam
Direct	−17.866			
Sequential 1	524.43	501.75	524.43	501.75
Sequential 2	−461.91	441.99	−461.91	441.99
Parallel 1	196.37	−213.79	1037.71	−1045.72
Parallel 2	196.37	−213.79	1037.71	−1045.72

From the mismatch numbers it becomes apparent that in both geometries the fifth-order signal is much better phase matched than either of the two cascaded processes. The other point of interest is that the crossed-beam geometry significantly affects the phase matching of the parallel cascade in comparison to the baseball geometry. In fact since the original experiments were deemed to be almost completely dominated by parallel cascades and free of sequential cascades [20] one might predict that the crossed-beam geometry should discriminate effectively against both cascaded signals, allowing the fifth-order signal to be measured.

Using an approach similar to that taken to obtain (1.7) an equation can also be arrived at for the cascaded signal [77]:

$$E_{\text{cas}}(\tau_2, \tau_4) = L^2 \frac{\omega_{\text{cas}}}{n_{\text{cas}}} \frac{\omega_{\text{int}}}{n_{\text{int}}} E^5 R_{\text{cas}}^{(3)}(\tau_2, \tau_4)$$

$$\times \frac{\sin(\Delta k_{\text{a}} L/2)}{\Delta k_{\text{a}} L/2} e^{i(\Delta k_{\text{a}} L/2)}$$

$$\times \frac{\sin(\Delta k_{\text{b}} L/2)}{\Delta k_{\text{b}} L/2} e^{i(\Delta k_{\text{b}} L/2)}. \tag{1.14}$$

This equation can be used for both the parallel and sequential cascades and takes as parameters the sample pathlength (L), the phase mismatch for the two steps in the cascade (Δk_{a} and Δk_{b}), the signal frequency ω_{cas} and index n_{cas}, and the intermediate signal frequency ω_{int} and index n_{int}. In comparison to (1.7) it should be noted that the dependence of the cascaded signal increases as L^2 whereas the fifth-order signal increases as L. This implies the longer the pathlength the worse the cascaded signal will become so one simple strategy for reducing cascades is to use short pathlength cells. Perhaps the most important feature of (1.14) is if the third-order time-dependent response is known then the cascaded signal can be completely simulated (Fig. 1.19). Generally speaking the third-order Raman response for simple liquids is well known and can be fit to analytical functions which can then be used in place of $R_{\text{cas}}^{(3)}$. To properly simulate the various possible tensor elements it is necessary to have the complete set of third-order polarization tensor elements. Third-order spectroscopy is represented by a fourth-rank polarization tensor of which only four elements are necessary to describe the third-order polarization in the sample [78]. Only three of these elements are independent

$$R_{1111}^{(3)}(t) = R_{1221}^{(3)}(t) + R_{1122}^{(3)}(t) + R_{1212}^{(3)}(t). \tag{1.15}$$

The response subscripts describe the relative polarizations of the various beams where 1 and 2 are orthogonal polarizations (sometimes also represented as Y and Z). By convention, the time ordering of the electric fields is represented in the subscript from right to left. Because pulses within a pulse pair are interchangeable in the nonresonant experiment, (1.15) reduces to

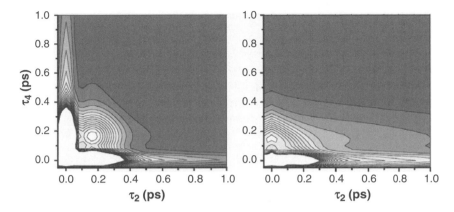

Fig. 1.19 Calculated perfectly phase-matched cascaded third-order signals for CS_2. The $\chi^{(3)}$ response of CS_2 is well characterized such that an analytical calculation of the cascaded signal can be made. Sequential cascade *(left)* and parallel cascade *(right)*. Note the signature common to both cascaded third-order signals is the appearance of signal along the pump delay, τ_2 axis ($\tau_4 = 0$). Also note the sequential and parallel cascades do not have exactly the same temporal dependence. The sequential cascade generates signal symmetrically about both axes; while the parallel process decays more quickly along the τ_4 ($\tau_2 = 0$) axis. Under poorly phase-matched conditions, the phase of the sequential cascade will be different from the parallel cascade. This effect and the different temporal dependences between the different cascaded channels can lead to nodes along τ_4 at approximately 200 fs, depending on the phase differences between the cascades and needs to be considered any time there is signal along the τ_2 fifth-order axis ($\tau_4 = 0$)

$$R^{(3)}_{1111}(t) = R^{(3)}_{1122}(t) + 2R^{(3)}_{1212}(t). \qquad (1.16)$$

There are thus only two independent third-order responses and all others can be constructed from those two. This implies one can construct all possible cascaded signals using only two measured signals, there are, however, a number of mixed tensor elements that are often directly measured. An example of this is the magic-angle signal which corresponds to

$$R^{(3)}_{11mm}(t) = 2R^{(3)}_{1122}(t) + R^{(3)}_{1111}(t), \qquad (1.17)$$

where m represents an angle of $54.7°$ with respect to the 1 polarization. This specific tensor element measures the isotropic part of the third-order response and has been predicted to be useful in analyzing the fifth-order response as well as in lowering the possible cascade contribution to the signal [79].

The advantages diffractive optics bring to nonlinear spectroscopy have been well established. By carefully designing the diffractive optic to diffract the beams into a pair of parabolic mirrors such that the above conditions are satisfied, we can greatly simplify our experimental setup as well as allow us to heterodyne detect our signal. Heterodyne detection is a particular advantage because for perfect phase-matching conditions, the fifth-order signal will be $\pi/2$

phase shifted from the cascaded signals [75, 80]. This should allow selection between the desired signal and the contaminant signal by adjusting the phase of the reference oscillator. In the case of mixed-phase signals, where there is no perfect $\pi/2$ phase shift between the two electric fields, it becomes impossible to clearly separate the two signals [23]. It is still possible to minimize or maximize the various signal components but complete isolation is no longer feasible and the subsequent analysis becomes more complicated.

1.4   Experimental Findings

Over the course of the past five years we have compiled an exhaustive set of experiments wherein we attempt to ensure we have measured and fully characterized the fifth-order Raman response of liquid CS_2. The most important results and conclusions are summarized here. In addition, preliminary results are presented for liquid benzene.

The initial experiments performed involved a homodyne signal comparison between the two geometries with a sample pathlength of 300 μm [11]. A two-color, diffractive optics-based approach was used: 400 nm pumps and 800 nm probe. To lessen the background scatter the probe polarization was rotated $+45°$ from the pumps, and the analyzer (signal) polarization was rotated 90° from the probe. The response measured was $R^{(5)}_{-45°\,45°\,1111}$. Comparing the results of the two phase-matching geometries (see Fig. 1.20), the two signals proved to be markedly different with the crossed-beam geometry lacking the strong

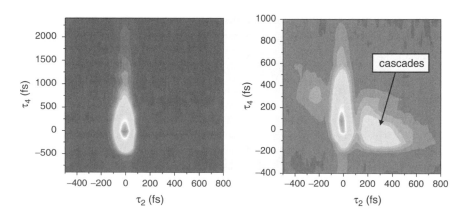

Fig. 1.20   Homodyne detected $R^{(5)}_{-45°\,+45°\,1111}$ fifth-order Raman signal of CS_2 measured using the crossed-beam *(left)* and baseball *(right)* geometries. Sample pathlength used was 300 μm. Note the disappearance of the cascade signature along τ_2 with the greater geometrical phase mismatch against the parallel cascade in the crossed-beam geometry. The signal along τ_4 remains unchanged. This comparison illustrated that the cascaded signal was negligible in the crossed-beam geometry [11]

ridge along the pump delay (τ_2) from the baseball geometry. Since this ridge is the signature of the parallel cascade [20] and since the primary difference between the two geometries is a large increase in phase mismatch for the parallel cascade (see Table 1.2), it was concluded that the crossed-beam geometry effectively discriminates against the parallel cascade. The remaining signal in the crossed-beam data was pulse-width limited along the pump delay (τ_2) and was very elongated along the probe delay (τ_4) with the signal persisting beyond 1 ps. No long-lived diagonal signal was visible, indicating a probable lack of inhomogeneous broadening. This description did not match the sequential cascade, the signature of which is signal symmetry along both axes, and contained no trace of parallel cascade (see Fig. 1.19). Furthermore, the data collected with the baseball geometry give a direct measurement of the signal amplitude originating from the parallel cascade contribution under phase-matched conditions. By simply exchanging spatial pinholes, the crossed-beam geometry was measured under identical conditions in which the phase matching discriminates against the parallel cascade to the same degree as the sequential cascades in both beam geometries. The signature of cascades is the signal along the τ_2 ($\tau_4 = 0$) axis, as indicated with an arrow in Fig. 1.20 for clarity. Fundamentally, there can be no fifth-order signal along this axis as there has not been enough time for any nuclear motion to respond to the excitation fields. As can be seen, this feature completely disappears when the phase matching is switched to the crossed-beam geometry that effectively discriminates against both parallel and sequential $\chi^{(3)}$ cascades. As such the data were presented as the first measurement of the homodyne fifth-order Raman response of CS_2 [11].

The next experiment took advantage of the diffractive-optic's built-in heterodyning capability and unblocked the reference to measure the phase and amplitude of the signal. The response measured was again $R^{(5)}_{-45°45°1111}$, the crossed-beam geometry was used, and the phase was cycled through a full π phase shift (see Fig. 1.21). The phase was assigned by setting the coverslip angle for the maximum positive signal as $\Delta\phi = 0$ and the maximum negative signal as $\Delta\phi = \pi$, all other phases are established by calibration of the rotation angle of the coverslip. Note that this method of setting the phase merely defines the relative phase at $\tau_2 = \tau_4 = 0$, this may not correspond to the maximum phase of the fifth-order nuclear signal (1.3). However by cycling through a full phase shift we ensure that all possible signals are measured and examined. The sample pathlength used for the pump delay scans (τ_2) was 300 μm and for the probe delay (τ_4) a sample pathlength of 100 μm was used. The sample pathlength was changed to help minimize the group-velocity mismatch along the probe delay: the two-color pump and probe pulses propagate through the sample at different speeds causing a smearing of the time resolution, making the pathlength shorter helps minimize this problem. This problem does not affect the time resolution

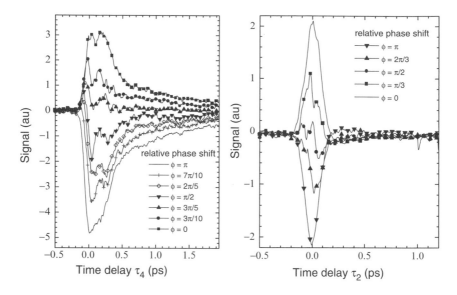

Fig. 1.21 Heterodyne-detected $R^{(5)}_{-45° +45° 1111}$ fifth-order Raman response of CS_2 measured using the crossed-beam geometry at various relative phase settings. The probe delay (*left*) was measured with a sample pathlength of $100\,\mu m$, the pump delay (*right*) was measured with a sample pathlength of $300\,\mu m$ [81]. Reprinted from [81]. Copyright 2000, with permission from Elsevier

along the pump delay where the probe pulse is not scanned. The heterodyne detection shows clearly that the signals along both axes are composed of only a single phase signal. There is no indication of any nodes or crossings which would be signatures of a second signal component of significant amplitude at a different phase. This appears to indicate that the only signal being measured is the true response and that the phase assignments used appear to be accurate. The fifth-order signal appears uncontaminated with any other signal component, and at the $\pi/2$ phase setting, where the cascaded signal is most likely to appear, the signal is negligible. This is the first heterodyne-detected fifth-order Raman signal of CS_2 and it appears to be free of any cascaded signal whatsoever [81].

Now that the signal had been established as measurable and uncontaminated with cascades, the next step was to fully characterize the response. Several papers had been published examining the merits of various polarization tensor elements and the potentially unique information they might contribute. Tokmakoff et al. had published a set of two papers examining the third-order [82] and fifth-order [83] tensor elements for the nonresonant Raman case and the dipole resonant case. The result was seven nonzero tensor elements that could be measured for the fifth-order Raman response: $R^{(5)}_{111111}$, $R^{(5)}_{112211}$, $R^{(5)}_{111122}$, $R^{(5)}_{221111}$, $R^{(5)}_{121211}$, $R^{(5)}_{111212}$, and $R^{(5)}_{121112}$. In addition to these signals there are a number of mixed tensor elements that can be measured where the relative polarizations of the pulses are not orthogonal. Perhaps the most important of these is the magic-angle signal. The magic-angle response is well known in the third-order experiment in which it isolates the isotropic response of the liquid. There is no simple way to separate the fifth-order response into isotropic and anisotropic responses but it had been predicted that $R^{(5)}_{11mm11}$ and $R^{(5)}_{1111mm}$ (where $m = 54.7°$) might help reveal the echo signal and eliminate portions of the diffusive tail [38, 79]. The seven independent fifth-order tensor elements were measured as well as $R^{(5)}_{11mm11}$, each was presented as a pump delay scan (τ_2), a probe delay scan (τ_4), and a diagonal scan ($\tau_2 = \tau_4$). Each was also presented with an in-phase trace and an in-quadrature trace. The CS_2 was flowed through a 100 μm pathlength cell. The dynamics of each data set is remarkably consistent no matter which tensor element was measured. The response characteristics match in every way the results from the previous experiments: a long-lived ridge along the probe delay, a pulse-width limited signal along the pump delay, and little signature of a rephasing echo along the diagonal (for an example see Fig. 1.22 which shows the fully polarized case). Intriguingly, the magic-angle response, $R^{(5)}_{11mm11}$, is remarkably similar to the fully polarized response, $R^{(5)}_{111111}$. There is the suggestion of a more pronounced signal along the diagonal but it is difficult to discern since it appears to be right on the edge of the pulse resolution.

There is the possibility that some of the signal may contain contributions from two-photon electronic transitions at the excitation levels and wavelengths used. To further test for this potential contribution, a novel three-color six-wave mixing experiment was executed using a noncollinear optical parametric amplifier to provide one of the excitation pulse pairs at a center wavelength of 530 nm [84]. The difference in excitation wavelength induces a significant change in the two-photon cross section such that the signal amplitude from any two-photon electronically resonant contributions should be dramatically affected. Under these conditions the signal amplitude and form were essentially identical to that of the two-color fifth-order response. This finding helped further rule out other contributions and further confirmed the signal purity.

Independent to the above work, Fleming and co-workers improved the signal to noise in their homodyne-detected experiment by improving the stability of their pump laser and introduced a new phase matching geometry that improved the contrast against the third-order cascades [85]. The observed signal response (see Fig. 1.23) did not match its calculated cascade signal response and it was argued that the signal was representative of the nuclear $\chi^{(5)}$ response. It was later shown that under nonphase-matched conditions the parallel and sequential cascades develop different phase factors in their respective contributions to the total signal field [23]. This phase difference leads to interference effects as the temporal profiles of the sequential and parallel cascades are not exactly the

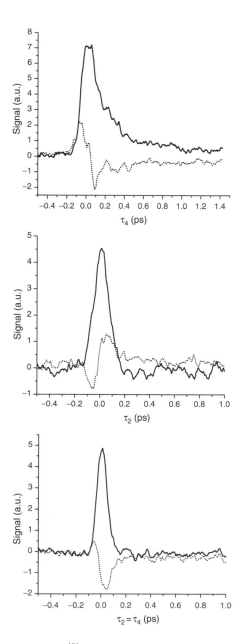

Fig. 1.22   Heterodyne-detected $R^{(5)}_{111111}$ fifth-order Raman response of CS_2 measured using the crossed-beam geometry and 100 μm pathlength. Two phase settings are shown: *solid line* is the in-phase signal ($\Delta\phi = 0$), *dotted line* is the in-quadrature signal ($\Delta\phi = \pi/2$). The signal scans are, *from top to bottom*, probe delay (τ_4), pump delay (τ_2), and diagonal scan ($\tau_2 = \tau_4$) [23]. Reused with permission from [23]. Copyright 2002, American Institute of Physics

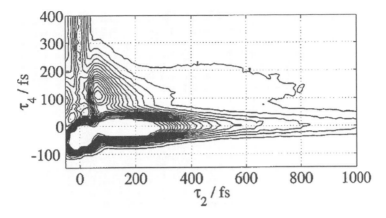

Fig. 1.23 Homodyne-detected $R^{(5)}_{111111}$ fifth-order Raman signal of CS_2 measured using one color at 800 nm and a 1 mm pathlength by Blank et al. [85]. Reused with permission from [85]. Copyright 2000, American Institute of Physics

same (see Fig. 1.19). At different points in the two dimensional time scan the amplitude of the sequential cascade (which decays more slowly along τ_4 and is asymmetric along the two axes) will partially or completely cancel the parallel cascade depending on the sample conditions and beam geometries. Using the reported beam and sample conditions, it was possible to simulate all of these data nearly quantitatively with cascades [23]. However, it was not possible to completely account for the off-diagonal features observed. To further improve the signal to noise and enable direct detection of the signal field, Kaufman et al. introduced an active-feedback loop on the reference and probe beams after the sample and obtained reasonable stability by careful mechanical stabilization of the pump beams [86]. These results are shown in Fig. 1.24. With the introduction of active-feedback heterodyne detection, several new features have appeared. First and most important is the observation of nodes along the probe delay. The emphasis of this work was on the close agreement of the position of nodes to the theoretical calculations [54] of Saito and Ohmine (see Fig. 1.13) for the different tensor elements studied. These nodes arise from differences in sign and temporal dependence of the nuclear anharmonicity and nonlinear electronic polarizability source terms in the nuclear $\chi^{(5)}$ response as discussed above. The nuclear anharmonicity terms contribute principally through the librational or hindered rotational contributions to the signal; these motions are strongly damped. Close inspection of the theoretically calculated response function and the experimental results shows that there are very significant discrepancies that need to be taken into account. The most significant is the strong signal along the pump delay axis (t_1 in Fig. 1.24). As discussed above, signal along this axis is generally considered the signature of lower order cascaded signal contributions. It was argued in this work that this large signal

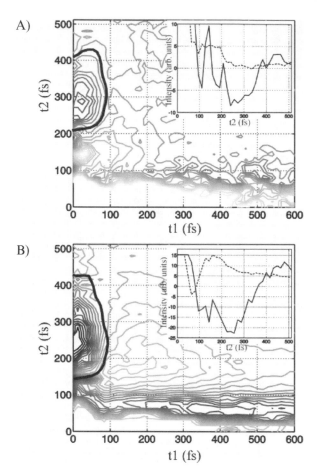

Fig. 1.24 Active feedback heterodyne-detected (**a**) $R^{(5)}_{1111mm}$ and (**b**) $R^{(5)}_{11mm11}$ fifth-order Raman signal of CS_2 measured by Kaufman et al. ($m = 54.7°$). The delay conventions used are t_1 = pump delay and t_2 = probe delay. *Dark lines* represent the approximate position of nodes in the signal. Sample pathlength used was 1 mm. Note the similarity in the nodal position along t_2 to that shown in the calculation of Fig. 1.13. The large signal along t_1 is not captured in the calculation and was attributed to electronic-nuclear hyperpolarizability contributions [86]. Reprinted with permission from [86]. Copyright 2002, American Physical Society

along the pump delay axis was due to electronic-nuclear hyperpolarizability contributions (see 1.18). This term results from a four-field interaction with the electronic-nuclear polarization. The problem with this explanation is that if there was such a strong contribution along this axis from this term, there would have to be a corresponding electronic-nuclear hyperpolarizability term contributing to the signal along the probe-axis ridge (t_2 in Fig. 1.24) where the nodes of interest occur and this term would certainly cancel or at least displace the position of the nodes from where they should be observed in the absence of this term. It should also be noted that the signal along both the pump and probe axes is at least an order of magnitude larger than any other contribution in the 2D plot. The signals along these axes have been truncated to enable observation of the key off-diagonal features. Given the magnitude of this signal, if it is due to the electronic-nuclear hyperpolarizability term, it would have to exactly vanish along the probe axis to truly reveal the nodes in the pure $\chi^{(5)}$ nuclear response. Taking into account the symmetry in the excitation and probe for this one-color experiment, this possibility seems very unlikely but cannot be completely ruled out without a full theoretical analysis of the electronic-nuclear hyperpolarizability term. Regardless, the off-diagonal features that were observed previously by this group near $t_1 = t_2 = 100$ fs (see Fig. 1.23) are still distinctive and represent the one feature in the measured signal where there is consistent agreement between both experiment and theory from different groups.

The major problem in the comparison of 2D fifth-order Raman spectra to theory is that the theoretical calculations do not include all the contributions to the experiment in their calculated response functions such that there is tendency to selectively isolate certain regions of the spectrum in which one finds coincidences. An additional means of verifying the purity and origin of the fifth-order signal was necessary. One of the most important contributions to the field was the discovery of a mixed-polarization tensor element by Jansen, Snidjers, and Duppen that leads to a factor of 10^4 increase in contrast against cascaded nuclear $\chi^{(3)}$ contributions (vide infra) [50]. The proposed response measured is $R^{(5)}_{60°60°11-60°-60°}$ where the polarizations are $60°$ counterclockwise for the first pump pair, $0°$ for the second pump pair, and $60°$ clockwise for the probe/analyzer. We have referred to this polarization setting as the "Dutch Cross" configuration in honor of its origin. This configuration provides essentially complete elimination of the cascades by taking advantage of the sign change between the $R^{(3)}_{1111}$ and $R^{(3)}_{1122}$ tensor elements. The different polarization angles effectively lead to cancellation of the third-order contribution. This high degree of additional contrast against the cascaded lower order responses provides the acid test so to speak with respect to the signal purity of the fifth-order response. The results of this experiment using the diffractive optics approach are shown in Fig. 1.25. Most of the same features appear: a long-lived signal along the probe delay and a pulse width limited signal along the pump delay. The most striking new feature is the signal along the diagonal where a pronounced shoulder has appeared extending beyond the pulsewidth and peaking around $\tau_2 = \tau_4 = 120-150$ fs. No other tensor element or polarization configurations so far have shown such a dramatic diagonal feature. This diagonal signal was largely masked by the much larger electronic hyperpolarizability signal at the time origin and overlap with the signal along the τ_4 ridge, most notably for the all parallel polarization configuration. The sign change of the nuclear response along the probe delay relative to the off-diagonal contributions more

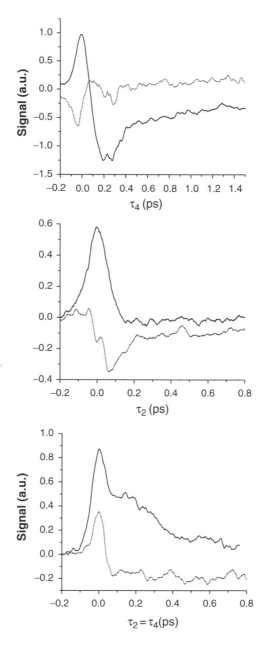

Fig. 1.25 Heterodyne-detected "Dutch Cross" fifth-order Raman response of CS_2 measured using the crossed-beam geometry and 100 µm pathlength. Two phase settings are shown: *solid line* is the in-phase signal ($\Delta\phi = 0$), *dotted line* is the in-quadrature signal ($\Delta\phi = \pi/2$). The signal scans are, *from top to bottom*, probe delay (τ_4), pump delay (τ_2), and diagonal scan ($\tau_2 = \tau_4$) [84]. The signal along the probe axis has a sign change that occurs within the pulse width. The nuclear rephasing response along the diagonal is clearly resolved. Reprinted from [84]. Copyright 2003, with permission from Elsevier

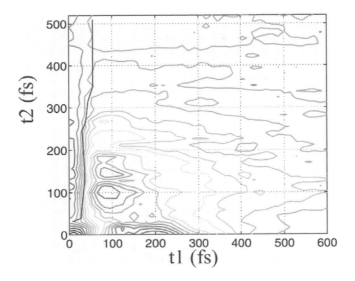

Fig. 1.26 Active feedback heterodyne-detected fifth-order Raman signal of CS_2 measured by Kaufman et al. using the Dutch Cross polarization settings. The node is represented by a *black line*. The delay conventions used are t_1 = pump delay and t_2 = probe delay. Sample pathlength used was 1 mm. Note the significant reduction in the signal along t_1 in comparison to Fig. 1.24 and the shift in the node along t_2 to less than 100 fs [87]. Figure courtesy of L.J. Kaufman and G.R. Fleming

than anything else now allows the diagonal signal to be clearly resolved. This study demonstrated that the previous diffractive optics experiments had indeed isolated the pure fifth-order nuclear response. This point is further reinforced by the most recent measurements of Kaufman et al. that are shown in Fig. 1.26 in which they performed the same "Dutch Cross" polarization configuration [87]. The most striking effect is the near elimination of signal along the pump axis (t_1 in Fig. 1.26). As emphasized above, signal along this axis is the clearest signature of cascades. The elimination of the signal along this direction with the "Dutch Cross" configuration indicates that there was a cascade contribution to the previous findings. Of much greater importance, however, is that the most pronounced off-diagonal features that could not be explained by cascades still remain.

Inspection of the different slices shown in Fig. 1.25 from the results of Kubarych et al. with that of Fig. 1.26 from the work of Kaufman et al. show that the two groups' results have now converged with the advent of the "Dutch Cross" configuration. The node position along the probe axis is approximately the same at <100 fs and there is agreement on the change in sign of the nuclear signal along the τ_4 ridge and the diagonal components. In fact the two experiments are in remarkable agreement as is further highlighted in Fig. 1.27 where slices along the diagonal from the two different studies are directly compared

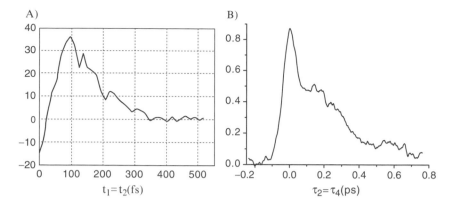

Fig. 1.27   Heterodyne-detected "Dutch Cross" fifth-order Raman signal of CS_2 measured by (**a**) Kaufman et al. and (**b**) Kubarych et al. [84,87]. *Note*: the electronic response at the time origin in relation to the nuclear response has a different sign in the one-color and two-color experiments. Taking this into account, the position of the maximum and the decay in the rephasing of the nuclear response are in very good agreement with this configuration. *Right figure* reprinted from [84]. Copyright 2003, with permission from Elsevier. The *left figure* was provided by L.J. Kaufman

for the same "Dutch Cross" polarization configuration. These results now agree extremely well with the various theoretical calculations for CS_2 that show a strong ridge along the probe axis, and peak in the diagonal or bath refocusing direction at approximately $120-150$ fs. After a long struggle, both the experiments and the theory are converging on the true nature of the fifth-order nuclear Raman response of CS_2. The only significant discrepancy is with the theoretical prediction for the node along the τ_4 axis. This feature of the signal is important as it provides a calibration for the degree of accuracy with which the relative anharmonic and nonlinear polarizability terms have been treated in the intermolecular interactions of the theoretical models. The results of Saito and Ohmine calculate the node to be at approximately 200 fs along the probe axis (see Fig. 1.13). These calculations were conducted for different tensor elements but the position of the node is largely insensitive to the tensor element in their work [55]. The work of Jansen et al. did not find a node in the finite field MD calculations (see Fig. 1.12). However, it has since been learned that an inaccurate compressibility for CS_2 was used in these calculations. We have since repeated these FFMD calculations [88] and we have been able to reproduce the calculations of Saito and Ohmine using the same DID model they used. These calculations are shown in Fig. 1.28 for the all parallel polarization response (see figure caption for details). This result now shows that the problem with the compressibility of CS_2 has been rectified and the two different means of calculating the fifth-order nuclear response function agree and converge to the same result. Also shown in Fig. 1.28 is the same calculation but using the

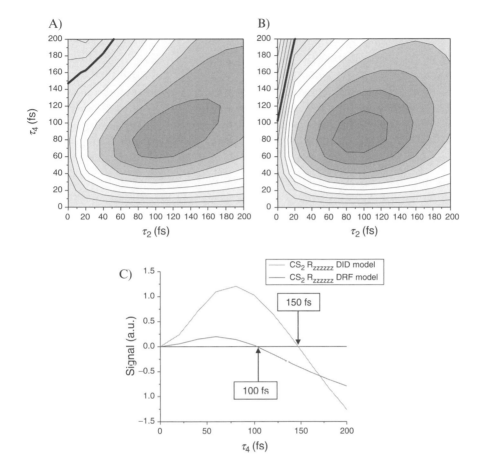

Fig. 1.28 FFMD calculations of the fifth-order nuclear response of CS_2. (**a**) DID model with corrected compressibility for the all parallel tensor element. These results are in excellent agreement with the work of Saito and Ohmine (see Fig. 1.13). (**b**) DRF model for the same tensor element. (**c**) Comparison of the node positions along the probe axis for the DID and DRF all parallel calculations. Note that the DID and DRF results are not normalized and comparison should only be made between overall dynamics and node positions. The simulations used 64 rigid CS_2 molecules and were averaged over 2,000 trajectories. The field strength used in both models was $1.829 \text{ V}/\text{Å}^{-1}$

DRF model. This model is a more accurate description of the intermolecular interactions as discussed previously. From the results it appears that the model used has a profound effect on the node position in the FFMD simulation. The results of Saito and Ohmine predict a node position around 200 fs for both DID and DRF calculations (see Fig. 1.13). In our FFMD simulation, both the DRF model and the DID model predict a much earlier node position, with the DRF model indicating a 50 fs shift to earlier time from the DID approximation (see

Fig. 1.28c). This preliminary result appears to indicate the FFMD calculated response is sensitive to the intermolecular interaction model chosen and that the node position varies with the model. The corollary to this is that the relative contributions of the anharmonic and nonlinear polarizability terms in the calculation are changing between the two models. As this change in sign along the probe axis is the one discrepancy between experiment and theory for this tensor element it remains an open question as to where the difference originates. Further calculations are in progress with a specific focus on the "Dutch Cross" tensor element where the experimental results have converged. The primary conclusion that should be drawn, however, is that the overall dynamics of the new simulations is in excellent agreement with previous MD calculations for the all parallel polarization response of CS_2. This convergence of both the theory and the experiment is an important milestone in the advancement of fifth-order Raman spectroscopy as a probe of the liquid state.

With the recent convergence in both experimental and theoretical understanding of the fifth-order nuclear response of CS_2, we now present some preliminary results in which the prospects for extending this form of spectroscopy to other liquids can best be appreciated. Liquid CS_2 has the advantage of an extremely large nonresonant nonlinear response relative to all other liquids; this factor unfortunately also contributes to the problem of isolating the signal from the very large third-order cascaded contaminant signal. As can be appreciated from the discussion of the background theory of the nuclear $\chi^{(5)}$ response, the magnitude of the fifth-order Raman response does not scale with the polarizability per se but rather depends on the nuclear anharmonicity and nonlinear dependence of the polarization on displacement (all second-order effects). Other liquids will likely have the same degree of nuclear anharmonicity and perhaps larger nonlinear dependences in the polarizability. It is not possible to state a priori what liquid would give a reasonable $\chi^{(5)}$ signal-based solely on the relative magnitude of the $\chi^{(3)}$ response of a particular liquid relative to that of CS_2. In fact, it is quite likely that other liquids may have comparable nuclear $\chi^{(5)}$ amplitudes with less problems with contamination from cascaded $\chi^{(3)}$ processes. The choice was made to try benzene and its derivative toluene. The original concept was to use the baseball geometry to obtain the cascaded third-order signal, which had been seen previously [59], and to use that to isolate the fifth-order signal using a combination of phase-control and the "Dutch Cross" geometry. In point of fact this was not necessary, though the signal is extremely low in benzene (about an order of magnitude down from CS_2), it is measurable using the crossed-beam geometry. The preliminary results for the response are shown in Fig. 1.29. The heterodyne signal was measured using a 500 μm pathlength cell which is longer than was used for CS_2 but necessary to obtain signal. The salient features along the axes are similar to those of CS_2: the pulse-width limited signal along the pump delay, a longer lived signal along the probe delay (signal persists past

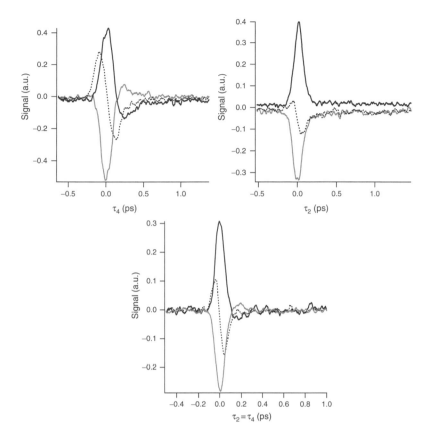

Fig. 1.29 Two-color heterodyne-detected fifth-order Raman response, $R^{(5)}_{111111}$, of benzene measured using the crossed-beam geometry and 500 μm pathlength. Three phase settings are shown: *solid line* is the in-phase signal ($\Delta\phi = 0$), *dotted line* is the in-quadrature signal ($\Delta\phi = \pi/2$), and *gray line* is the out-of-phase signal ($\Delta\phi = \pi$). The signal scans are, *clockwise from topleft*, probe delay (τ_4), pump delay (τ_2), and diagonal scan ($\tau_2 = \tau_4$). Note the small amplitude nuclear signal along the diagonal at approximately $\tau_2 = \tau_4 = 100$–200 fs and along τ_4; there is no discernible signal along the τ_2 axis to indicate cascade contributions

500 fs) with a change of phase at around 200 fs. The diagonal signal is more interesting since there appears to be signal well outside the pulse width, persisting to ∼400 fs. Work is continuing on these experiments with the goal of achieving stronger signal so as to move to shorter pathlengths and measure other tensor elements. Specifically the "Dutch Cross" geometry still needs to be examined as it has become necessary for confirming any fifth-order signal. In addition, we have conducted finite field MD calculations for the fifth-order response of benzene and find reasonable agreement with the experimental results to date. Both aspects of this work indicate that moderate improvements in laser technology will enable a much broader extension of this new form of spectroscopy to more complex liquids as elaborated below.

1.5   General Discussion and Implications of the Findings for the Liquid State

As has been clearly demonstrated this is a difficult experiment from the point of view of characterizing the signal. The discussion so far has centered on the possible cascaded signals and their contamination of the fifth-order response but there is another contribution which must also be taken into account. There are four possible terms that can appear in the fifth-order response [89]:

$$R^{(5)}(\tau_2, \tau_4) = \langle \zeta \rangle \delta(\tau_2)\delta(\tau_4) + \frac{i}{4\hbar}\langle[\alpha(\tau_4), \gamma(0)]\rangle\delta(\tau_2)$$
$$+ \frac{i}{2\hbar}\langle[\gamma(\tau_2), \alpha(0)]\rangle\delta(\tau_4) - \frac{i}{4\hbar^2}\langle[[\alpha(\tau_2 + \tau_4), \alpha(\tau_2)]\alpha(0)]\rangle.$$
$$(1.18)$$

The first term corresponds to the nonresonant electronic hyperpolarizability which only exists at the time origin and is generally pulse-width limited, the second and third terms correspond to a electronic-nuclear hyperpolarizabilities along the axes and the last term is the proper nuclear fifth-order response. From the above equation it is clear that some careful signal analysis is necessary to separate the various signals along the axes and ensure any conclusions drawn apply to the correct measurement. Fortunately each term is affected differently depending on which polarization configuration is used. From its initial use to measure the homodyne signal from CS_2, the crossed-beam phase-matching geometry coupled with short pathlengths has proven singularly adept at measuring the uncontaminated fifth-order Raman response. The data consistently display the same features with few discrepancies through all the measured tensor elements, through all the different experiments. Perhaps most tellingly the signal remains consistent through the "Dutch Cross" polarization settings. The fifth-order response for this signal is

$$R^{(5)}_{60°60°11-60°-60°}(t) = \frac{3}{8}\left(R^{(5)}_{221111}(t) + R^{(5)}_{112211}(t) + R^{(5)}_{111122}(t)\right)$$
$$- \frac{1}{8}R^{(5)}_{111111}(t).$$
$$(1.19)$$

At first glance this appears to be an odd choice since the above expression contains a combination of four different tensor elements and does not appear to contain any intuitive meaning. The key to the tensor element lies in the cascaded signals generated by this response. The following expression gives the third-order response for a mixed polarization state [50]:

$$R^{(3)}_{11\theta\theta}(t) = \cos^2\theta R^{(3)}_{1111}(t) + \sin^2\theta R^{(3)}_{1122}(t),$$
$$(1.20)$$

where θ is the polarization angle of the pumps with respect to the probe polarization. When an angle of $60°$ is used the resulting third-order signal has a strongly suppressed nuclear response as well as a reduced electronic response. The result is a signal which consists of solely an electronic hyperpolarizability at the time origin and little to no signal at later times. The cascaded signals that result from this would also be limited to the time origin with little contribution at times outside the pulse duration. This would effectively remove the cascaded signal from the fifth-order signal allowing the fifth-order nuclear response (1.3) to be seen. For an experiment measuring solely cascaded third-order contaminants the effect of switching to this polarization would most likely completely eliminate the signal, or at the very least change its shape drastically. Switching to this polarization setting had little impact on our signal, and more importantly the dynamics of our signal remains unaffected. A sign change now occurs along the probe axis but it appears to be within our pulse width making it likely corresponds to the electronic hyperpolarizability mixing with some longer lived signal with a different phase.

The pulse-width limited response along the pump delay is completely consistent through all the CS_2 experiments. The signal along this axis appears to be entirely limited to the nonresonant electronic hyperpolarizability response with no nuclear signal appearing at times longer than that expected for a pulse-width limited response (i.e., along $\tau_2; \tau_4 = 0$). This hyperpolarizability is generally considered an artifact that corresponds to all five ultrashort laser pulses being overlapped in time and space in the sample. A similar feature is seen in third-order nonresonant transient-grating studies where the three pulses create a similar signal at the time origin [21]. Though this is a fifth-order signal that is dependent on all laser pulses, there is no useful physical information contained within it. Its only use is it generally represents a crosscorrelation of the laser pulses in the sample and thus reflects the pulse widths. Outside of this signal there is nothing along the pump axis. There is not sufficient time for any nuclear motion that could contribute to pure the nuclear $\chi^{(5)}$ signal contribution of interest along this axis and only the aforementioned electronic-nuclear hyperpolarizability term can generate signal along this axis that is not related to cascades. Since such terms can be added ad hoc to explain fifth-order data, there need to be additional checks to rule out cascades and it is here that the "Dutch Cross" configuration figures prominently. With this polarization configuration, there appears to be no contribution along this axis from the electronic-nuclear hyperpolarizability term, $\langle [\gamma(\tau_2), \alpha(0)] \rangle$, from (1.18). Furthermore, the relative magnitudes of the peak in the nuclear response along the diagonal ($\tau_2 = \tau_4 = 120$–$150\,\mathrm{fs}$) and that along the τ_4 axis ($\tau_2 = 0$) are very close to the calculated responses such that there appears to be very little signal contribution from this electronic-nuclear hyperpolarizability term. It might be expected to be visible in tensor elements where the second pump pair and the probe/analyzer polarizations are all parallel but this also does not appear to be the case by comparing signal magnitudes.

The long-lived signal along the probe delay in CS_2 is also completely consistent from the homodyne data through to the three-color experiment. The signal generally rises quickly to a maximum at the time origin, corresponding to the leading edge of the hyperpolarizability, then decays slowly to zero over the span of around 1 ps. The signal along this axis for the majority of the tensor elements decays in a very similar fashion to the nuclear *free-induction decay* that appears in the third-order signal. The dynamics along this axis is predicted to carry information about the population relaxation of the system. Steffen and Duppen performed an analysis [90] of the fifth-order response to show that in principle the population relaxation and dephasing contributions to the signal decay can be separated. After assessing all the various polarization tensor elements and their expected contributions to the electronic-nuclear hyperpolarizability term, $\langle [\alpha(\tau_4), \gamma(0)] \rangle$, along the probe axis the conclusion drawn was that this signal is probably indicative of population relaxation. The primary difficulty with assigning this specific information to the decay of the signal along the probe axis is that it remains unclear how much to attribute to the electronic-nuclear hyperpolarizability term. This signal contribution strictly involves a single time variable and is identical in nuclear response to $\chi^{(3)}$ Raman signals and as such would contain no new information beyond that obtained with $\chi^{(3)}$ experiments along this axis. The corresponding term along the pump axis is zero but no estimates have been made as to their predicted magnitudes or whether to expect differing magnitudes for each axis. However, as noted above, the comparative amplitudes for the off-diagonal features and this axis with theoretical predictions that only incorporate the fifth-order Raman response or nuclear $\chi^{(5)}$ source terms suggest there is very little electronic-nuclear hyperpolarizability contributions that are single time variable or one-dimensional signal contributions. The relative magnitude of this electronic-nuclear hyperpolarizability to the pure nuclear $\chi^{(5)}$ response is still an open question. With this qualification, the signal can be interpreted in accordance with the analysis of Steffen and Duppen in which case the strong similarity to the decay along this axis indicates that the liquid modes of CS_2 are strongly damped and the liquid is effectively homogeneously broadened. It is possible to specifically structure the excitation pulse bandwidth to excite specific frequency components so that it may be possible to directly determine the lifetimes of excited modes from the signal along the probe delay axis.

The only signal component of CS_2 remaining to be discussed is along the time diagonal. This is the portion of the fifth-order response that was originally anticipated would give the most information. It is this aspect of the fifth-order Raman signal that gives a direct window onto the bath echo or ability to rephase liquid state modes. As stated previously, it is an intriguing concept that within all the stochastic fluctuations within a liquid it would be possible to rephase the different motions on sufficiently short timescales. The dynamics of the rephasing process casts out the correlations that must exist within the relative motions of molecules in the liquid state. The liquid state involves large amplitude anharmonic motions of molecules at densities comparable to solids. In order for one molecule to undergo such motions, other molecules in the immediate vicinity must similarly move and there will be net correlations imposed on these motions until the motions are damped or phase relationships destroyed through collisions with uncorrelated density fluctuations. Is it really possible to rephase a liquid in this context? The results indicate that it is possible to rephase the bath modes but only on timescales corresponding to a period or less of the highest frequency librational/hindered rotations and translation-like motions of CS_2, in other words on the timescale of periods corresponding to the high frequency cutoff of the spectral density of states for CS_2 at around 100 cm^{-1}. For CS_2 there is no clear separation of hindered rotation/librations and translation motions [48, 91]. The rephasing process is most readily observed in the "Dutch Cross" polarization setting, where two independent experiments now show it to have a maximum between 120 and 150 fs, and the rephasing quickly decays to zero within 400 fs. This dynamics corresponds to strongly damped frequencies within 200–50 cm^{-1}. Based on this evidence, liquid CS_2 does not have much rephasing ability and one can infer extremely fast loss in correlations due to rapid energy exchange and elastic changes in momentum that lead to memory loss. Any collective basis used to describe these motions on short timescales would quickly collapse to motions better represented as single particle uncorrelated motions and the onset of diffusive behavior. Individual molecules have little sense of their surrounding neighbors due to the extremely rapid loss in correlations. This is perhaps what might be expected of this liquid but it is the experiment which confirms this viewpoint. The dynamics of liquid CS_2 appears to fall toward the homogeneous broadening limit from which we can conclude that CS_2 behaves as a simple liquid on essentially all relevant timescales.

The most encouraging result that can be drawn from the collection of CS_2 data is its convergence with the theoretical results. Equilibrium and nonequilibrium MD simulations have been performed by Saito and Ohmine [54, 55], and FFMD simulations by Jansen, Snijders, and Duppen [49,50,52,53]. The FFMD results are clearly in line with the experiment. The DRF model appears to have captured all of the dynamics observed quite well (see Fig. 1.28): the elongated response along the probe delay, the truncated response along the pump delay and the lack of long-lived echo along the diagonal. Most importantly the degree of agreement between the experiment and theory depends very strongly on the interaction model used in the calculation. In this regard, the work of Saito and Ohmine is particularly instructive. They were the first to observe a node or sign change between the signal along the probe delay and the off-diagonal components. They showed that this nodal plane arises from interference effects between the nuclear anharmonicity (anharmonic coupling between the rotational and vibrational/hindered translational degrees of freedom) and the nonlinear polarizability source terms. As such, this experiment is exquisitely sensitive to the two most important factors dictating intermolecular interactions. In order for the theory to properly predict the two-dimensional fifth-order Raman spectrum, the absolute magnitudes of the nuclear anharmonicity and coordinate-dependent nonlinear polarizability must be accurately treated. The specific shape along any given slice in the 2D plot may seem no more distinctive than 1D experiments, however the full 2D spectrum is an exceptionally sensitive probe of the many-body interactions in liquids. At the moment, it appears that the DRF model is required to obtain reasonable agreement, which indicates electron overlap plays an important role in the intermolecular interactions of CS_2. More recent simulations [55] by Saito and Ohmine have used the DRF interaction model and show the same behavior with nodes appearing in most of the tensor elements at approximately 200 fs or in the same location as their earlier DID calculations [54]. The experimental results show that this node appears at less than 100 fs but only in the $R^{(5)}_{221111}$ and "Dutch Cross" tensor elements, all other tensor elements show no node along the probe axis (see Figs. 1.22 and 1.25). The DRF model used in these calculations is slightly different from the one used to calculate the 2D response shown in Fig. 1.28 that was optimized to treat the CS_2 dimer [52]. The necessary next calculation is the "Dutch Cross" response which has proven to be the intersection point for the experimental results (see Fig. 1.27). The noted differences at this level of treatment of the intermolecular interactions further calibrate us to the sensitivity of the experiment and the impact it will make in refining our understanding of liquid state dynamics.

In relation to other theoretical treatments, there has clearly been a convergence in the predicted 2D fifth-order Raman response. Of the various theoretical methods described in Sect. 1.2, an overwhelming number predicts the probe delay ridge as the signature for the fifth-order response. From the initial MD results of Ma and Stratt for liquid Xe through to the most recent results of Keyes, Space et al. for CS_2, the response shows qualitatively similar trends in the predicted dynamics. Now that the models have been established as being consistent with both experiment and the various analytical approaches some new insight should emerge as the basic physics of the nonlinear interaction and connection to the liquid state is in hand. The analytical approaches aid this evolution in our understanding. For example, the adiabatic INM theory successfully duplicated the behavior of the MD simulation for liquid Xe when it took into account the anharmonicity of the potential (see Figs. 1.8 and 1.10). When the model relied solely on the nonlinearity of the polarizability it predicted a nonexistent echo signal (see Fig. 1.7). Confirmation of this can be seen in the work of Tanimura where his calculations involving only linear polarizability and a strongly damped anharmonic potential (see Fig. 1.5) are the closest to the experimental results. On this basis, we can conclude that the dominant signal source for liquid CS_2 stems from the anharmonicity of the intermolecular modes. Whether this is the case for other liquids remains to be seen. With respect to the rapid loss in rephasing ability of the liquid that has been predicted by most theories, *mode mixing through polarizability* as discussed by Saito and Ohmine likely plays an important role. However, in their INM analysis of this effect, the echo was not completely suppressed as observed experimentally. Again, the experimental observations of fast and complete loss in bath rephasing are consistent with the signal being dominated by nuclear anharmonicity in which case this rapid decay is due to rapid energy exchange between the different correlated motions and the associated loss of memory through collisions. All of this dynamics occurring within what could be described as near the single collision limit, the period of hindered translational motions, within a discrete molecular basis to describe the motions. In this context, the original assumption used by Tanimura and Mukamel to analytically derive the rephasing processes at the fifth-order of the field interaction is not strictly valid. This theoretical treatment assumed harmonic multimode Brownian oscillators to describe the liquid; it was well recognized that this approximation would only even approximately hold at very short times. The more important assumption that needs reevaluation is that of level independent dephasing in the derivation. As can be seen by the fifth-order Raman experiments, liquid CS_2 loses memory on very short timescales. Uncorrelated diffusive motions occur on a 100 fs timescale. This timescale is close to a single period of even the highest frequency components in the intermolecular spectral density. This being the case, it is clear that the liquid is marginally stable. The concept of a mode to describe the ground state to excited state transition of the induced polarization as a 0–1 one-quantum transition is correspondingly marginally correct. The potential energy landscape of the liquid must have very small barriers separating the different nuclear configurations to enable motions and this landscape is constantly dynamically evolving. The overtone of any modal description of the liquid dynamics would be unbounded by such a potential and rapid fluctuations: the two-quantum transition would have much stronger damping than the corresponding fundamental. The assumption of level-independent dephasing in liquids is not strictly valid. With this recognition, the very rapid decay in the rephasing ability within the fifth-order response is largely due to the damping of the Raman overtone that would dominate the dephasing. This effect may seem to mitigate the primary motivation for this experiment, however, this feature means the signal along the diagonal directly accesses the anharmonic nature of the liquid state. Furthermore, one should not expect a single time slice, no matter along which time profile, to sufficiently gain access to new information over conventional 1D linear spectroscopies. One needs some perspective of the entire surface of the

2D spectrum. It is very clear from the above discussion that the 2D spectrum is exquisitely sensitive to the major physical processes that govern the highly anharmonic motions of molecules in the liquid state.

The preliminary results of benzene (see Fig. 1.29) are primarily exciting because this is the first liquid besides CS_2 to have divulged its fifth-order response. Additionally signal has been observed from toluene as well and the combination of these two results give hope for the future of the spectroscopy for other liquids. Overall the signal measured from benzene shows the same characteristic features as CS_2: the probe delay ridge, the lack of signal along the pump delay, and the indication of a brief signal along the diagonal. The dynamical details are, however, slightly different. The first important distinction is that the ridge in benzene is negative, indicating, in all probability, a phase change between the electronic hyperpolarizability and the nuclear signal. This behavior is not seen in the all parallel tensor element in CS_2. The second distinction is that the probe delay ridge in benzene decays to zero within 600 fs. In CS_2 this signal decays to zero after 1 ps and is much longer lived. If this signal corresponds to population relaxation, as we have attributed it, then the relaxation is occurring much more quickly in benzene than in CS_2. If this signal corresponds to the electronic-nuclear hyperpolarizability term then the overall magnitude and dynamics of it are quite different between the two liquids. One might predict, based on their third-order response, that the two liquids should have similar contributions from this term. If this is indeed the case then it appears likely that the signal decay along the probe delay represents population relaxation. Further tensor elements must be measured for benzene before a conclusion can be definitively drawn. The diagonal feature in benzene is again different from its counterpart in CS_2. The sign of the echo in benzene is negative with respect to the electronic term at the time origin, no sign change is seen in CS_2. The benzene echo has no evident rephasing maximum outside the pulse width and decays to zero within 300 fs. The echo signal seen in CS_2 is most prominent in the "Dutch Cross" polarization configuration, this still needs to be measured in benzene where perhaps further dynamics will be revealed.

1.6   Summary and Future Perspective

In its short infancy, fifth-order Raman spectroscopy has caused a remarkable shift in our theoretical and experimental understanding of liquids. Theories have been raised up and cast down and experiments have been exhaustively characterized and criticized all to ensure that we properly understand these low-frequency dynamics. In some sense, the fifth-order Raman response has served as a "proving ground" for both theory and experiment. The 2D fifth-order Raman experiment represents an extraordinary challenge to the experimentalist. The signal is inherently very small and new methodologies were required to be able to pull out the signal from background contamination with sufficient signal to noise to enable a reasonable analysis and execution of control studies. The contributions from all the various groups [11,20,23,29,58–67,69,75,77,81,84–86,92] involved in this work have made a very significant contribution to the field of nonlinear optics in general. We now have a much better appreciation for the complex nonlinear interactions that occur at high orders of the electric field. Basically, all possible nonlinear interactions that can occur, will occur, and specific phase matching geometries were developed to isolate the high-order signals of interest. The ability to generate phase-locked pulse pairs, direct detection of the signal field, and the explicit use of phase to analyze signals have been advanced through the pursuit of this spectroscopy. Thanks to these efforts we are now that much closer to the use of complex pulse sequences and phase cycling as performed in NMR to extract signal responses of interest from complex systems. These developments will certainly have a very significant impact in other multidimensional spectroscopies such as 2D IR [73,93] and electronic spectroscopies [94]. Theoretical advances have likewise been made both in the level of sophistication of treating the intermolecular interactions and in the analytical formalism for handling the multitime correlations functions of liquids. These advances are already beginning to impact in other areas of spectroscopy [95]. It should be noted that the high sensitivity of the fifth-order Raman response to the details of the interaction potential, as determined from the theoretical work, stands alone as a very important singular development. The important feedback mechanism between experimental observation and theory in the refinement of our treatment of the many-body potential of liquids must involve an observable that is highly sensitive to the anharmonic terms in the potential and polarization effects. The comparison of theory to experiment for 1D spectroscopies, or even structural probes such as neutron and X-ray scattering of liquids, shows the observable is found to be relatively insensitive to the details of the assumed basis. Using water as the classic example, 1D Raman [96], IR [97], and reduced density functions derived from X-ray [98] and neutron scattering [99] can be fit well with a number of different bases for water. These observables probe dynamics that involve linear response and/or very small fluctuations near minima in which harmonic terms of the interaction potential dominate. Models that get the density and linear response terms correct will give very similar agreement. The key distinction of the liquid state, however, is the highly anharmonic nature of the molecular motions defining this equation of state. The fifth-order Raman response directly accesses these motions and gives us perhaps our most sensitive probe of liquids to date.

The field of fifth-order Raman spectroscopy is still in its infancy. There are many experimental and theoretical challenges that lie ahead. The simple liquid CS_2 has finally been resolved and there is preliminary evidence that this form of spectroscopy can be extended to other systems with signal having been seen from benzene and toluene. It is still increasingly important to apply all of the checks and balances developed over the course of these past few years to ensure any future results are trustworthy. We believe the use of diffractive optics has greatly advanced this experiment and will play an important role in future versions of the experiment. A well-characterized phase-matching geometry and short sample pathlengths are also important to keeping the cascaded third-order signals suppressed. Heterodyne-detection is also crucial since without the phase information about the various signal components, too much information is lost to properly analyze the fifth-order response. The "Dutch Cross" polarization configuration must always be one of the tensor elements measured to check for cascade contamination. This neatly summarizes the criteria that need to be met by future experiments but perhaps the most important point of all to be made is that these results tie in so tightly with theory that the two need to advance together. Little progress would have been made in either field without its counterpart pushing it on.

Now that benzene has appeared on the horizon as the next liquid to be scrutinized by the fifth-order Raman community the future looks bright. The next important challenge will be to measure the 2D response from a hydrogen-bonded liquid. Such liquids are much more strongly associated and should have longer lived correlation functions than CS_2 that can be used to determine the degree to which mode mixing through polarizability suppresses the echo signal. Experimental and theoretical means need to be developed to separate the electronic-nuclear hyperpolarizability contribution from the signal if nodes in the 2D spectrum and information on population relaxation are to be confidently extracted from the probe delay axis. We have suggested using the relative magnitudes of the off-diagonal components to those along the probe axis in comparison to the calculated response with this term not included. A detailed theoretical treatment of the relative signal magnitudes from the different source terms in the fifth-order response is clearly needed.

The ultimate goal is undoubtedly still to measure the fifth-order response of water [30]. We will never come to a complete understanding of the anomalous properties of water and its connection to living systems until we have a direct experimental determination of its many-body potential. The fifth-order Raman experiment provides a means of probing collective motions involving large enough number of water molecules to approximate effects out to the second shell and further. The unique properties of water are undoubtedly connected to the hydrogen bond network that imposes both correlations and rapid energy exchange between the different degrees of freedom [73]. The result is a highly associated liquid of low viscosity, two properties usually diametrically opposed. By combining 2D IR to probe the first shell of the intermolecular water correlations and 2D Raman to access longer range correlations, we should be able to directly determine the couplings between water accurately enough to provide sufficient details of the many-body potential of liquid water. This information allows us to fully appreciate its unique properties in microscopic terms. The main challenge is to once again further improve the sensitivity of the fifth-order Raman experiment. Calculations of the fifth-order Raman signal from water relative to CS_2 find that the signal will be approximately 600 times smaller [54]. This tiny level of signal is indeed a challenge and it may remain one of the great future experiments for some years to come. However, the prospect of determining the fifth-order Raman response of water is at least possible in principle, there are no fundamental limitations impeding its measurement. The experiment needs to be improved by a factor of 500 by increasing the signal amplitude and detection sensitivity. The experiments performed on CS_2 used pulses that were on the order of 100 fs. To access the two-quantum transitions needed for the rephasing step in the pulse protocols, one needs at least 2,000 cm^{-1} of bandwidth to study liquid water which corresponds to <10 fs pulses. Such laser systems with millijoule pulses are now available. Using heterodyne detection, the signal scales according to $I^{2.5}$ power for the various pulses such that the increased peak power, through the use of shorter pulses alone, could provide the needed increase in signal. The main limitation here will be dielectric breakdown in defining the maximum power density that can be used for these experiments. Water has a much larger threshold for dielectric breakdown than CS_2 such that it should support more intense focusing conditions allowing for a corresponding increase in signal. In addition, improvements in reducing laser noise and better phase stabilization will have a very dramatic effect on the ability to increase the detection sensitivity. It was not possible to fully increase the amplitude of the reference field that served as the local oscillator in the detection process beyond a factor of 10–100 gain due to laser noise on the reference and phase stability limitations. One can readily anticipate factors of more than ten improvements can be made here. Thus, at least in principle, the fifth-order Raman experiment of liquid water is within reach. This experiment will greatly test the limits of laser technology, signal detection, and theory. It is at the limits of what is currently possible but the destination is worth the voyage and there will be many liquids to visit along the way.

Acknowledgments

We thank Professor D.A. Reichman, Professor Y. Tanimura, Professor B. Space, Professor R.M. Stratt, Dr. T.l.C. Jansen, and Professor S. Saito for allowing us to include their work in this effort. Special thanks are extended to Dr. T.l.C. Jansen for his help in making the FFMD calculations work. We also thank Professor L.J. Kaufman and Professor G.R. Fleming for sending us their data and giving us permission to publish it. With your collective help we believe this chapter represents an important milestone in the nonlinear spectroscopy of liquids.

References

[1] Enderby, J.E. (1983). Neutron-scattering and the structure of liquids. *Contemp. Phys.* 24:561–575; Crozier, E.D., Rehr, J.J., Ingalls, R. (1988). *X-Ray Absorption: Principles, Applications, Techniques of EXAFS, SEXAFS, and XANES*, edited by Koningsberger, D.C., Prins, R.,Wiley, New York; Kusalik, P.G., Svishchev, I.M. (1994). The spatial structure of liquid water. *Science* 265:1219–1221.

[2] Oliver, W.F., Herbst, C.A., Wolf, G.H. (1991). Viscous liquids and glasses under high pressure. *J. Non-Cryst. Solid* 131–133:84–87; Granato, A.V. (1996). The shear modulus of liquids. *J. Phys. IV* 6:1–9.

[3] Palese, S., Schilling, L., Miller, R.J.D., Staver, P.R., Lotshaw, W.T. (1994). Femtosecond optical Kerr effect studies of water. *J. Phys. Chem.* 98:6308–6316.

[4] Frenkel, I. (1955). *Kinetic Theory of Liquids*, Dover, New York; Seeley G., Keyes, T. (1989). Normal-mode analysis of liquid-state dynamics. *J. Chem. Phys.* 91:5581–5586.

[5] Lanzafame, J.M., Palese, S., Wang, D., Miller, R.J.D., and Muenter, A.A. (1994). Ultrafast nonlinear optical studies of surface reaction dynamics: Mapping the electron trajectory. *J. Phys. Chem.* 98:11020–11033.

[6] Miller, R.J.D., McLendon, G., Nozik, A., Schmickler, W., Willig, F. (1995). *Surface Electron Transfer Processes*, VCH publishers Inc. New York; Bixon, M., Jortner, J. (1993). Solvent relaxation dynamics and electron-transfer. *Chem. Phys.* 176:467–481.

[7] Koper, M.T.M., Mohr, J.H., Schmickler, W. (1997). Quantum effects in adiabatic electrochemical electron-transfer reactions. *Chem. Phys.* 220:95–114.

[8] Castner, E.W., Chang, Y.J., Chu, Y.C., Walrafen, G.E. (1995). The intermolecular dynamics of liquid water. *J. Chem. Phys.* 102:653–659.

[9] Lee, H.W.H., Patterson, F.G., Olson, R.W., Wiersma, D.A., Fayer, M.D. (1982) Temperature-dependent dephasing of delocalized dimer states of pentacene in p-terphenyl: Picosecond photon echo experiments. *Chem. Phys. Lett.* 90:172–179.

[10] Muller, L.J., Vandenbout, D., Berg, M. (1993). Broadening of vibrational lines by attractive forces: Ultrafast Raman echo experiments in a $CHCl_3{:}CDCl_3$ mixture. *J. Chem. Phys.* 99:810–819.

[11] Astinov, V., Kubarych, K.J., Milne, C.J., Miller, R.J.D. (2000). Diffractive optics implementation of six-wave mixing. *Opt. Lett.* 25:853–855.

[12] Tokmakoff, A., Fayer, M.D. (1995). Homogeneous vibrational dynamics and inhomogeneous broadening in glass-forming liquids: Infrared

photon echo experiments from room temperature to 10 K. *J. Chem. Phys.* 103:2810–2819; Hamm, P., Lim, M., Hochstrasser, R.M. (1998). Non-Markovian dynamics of the vibrations of ions in water from femtosecond infrared three-pulse photon echoes. *Phys. Rev. Lett.* 81:5326–5329.

[13] Spencer, C.F., Loring, R.F. (1996). Dephasing of a solvated two-level system: A semiclassical approach for parallel computing. *J. Chem. Phys.* 105:6596–6606; Loring, R.F., Yan, Y.Y., Mukamel, S. (1987). Time- and frequency-resolved fluorescence line shapes as a probe of solvation dynamics. *Chem. Phys. Lett.* 135:23–29.

[14] Stratt, R.M., Maroncelli, M. (1996). Nonreactive dynamics in solution: The emerging molecular view of solvation dynamics and vibrational relaxation. *J. Phys. Chem.* 100:12981–12996.

[15] Palese, S., Miller, R.J.D. (1994). Unpublished; preliminary report based on interference with $\chi^{(3)}$ signal, Ultrafast Phenomena IX.

[16] Tanimura, Y., Mukamel, S. (1993). Two-dimensional femtosecond vibrational spectroscopy of liquids. *J. Chem .Phys.* 99:9496–9511.

[17] Bout, D.V., Berg, M. (1995). Ultrafast Raman echo experiments in liquids. *J. Raman Spectr.* 26:503–511.

[18] Wright, J.C., Ivanecky, J.E. (1993). An investigation of the origins and efficiencies of higher-order nonlinear spectroscopic processes. *Chem. Phys. Lett.* 206:437–444.

[19] Ulness, D.J., Kirkwood, J.C., Albrecht, A.C. (1998). Competitive events in the fifth order time resolved coherent Raman scattering: Direct versus sequential processes. *J. Chem. Phys.* 108:3897–3902.

[20] Blank, D.A., Kaufman, L.J., Fleming, G.R. (1999). Fifth-order two-dimensional Raman spectra of CS_2 are dominated by third-order cascades. *J. Chem. Phys.* 111:3105–3114.

[21] Goodno, G.D., Dadusc, G., Miller, R.J.D. (1998). Ultrafast heterodyne-detected transient-grating spectroscopy using diffractive optics. *J. Opt. Soc. Am. B* 15:1791–1794.

[22] Dadusc, G., Goodno, G.D., Chiu, H.L., Ogilvie, J., Miller, R.J.D. (1998). Advances in grating-based photoacoustic spectroscopy for the study of protein dynamics. *Isr. J. Chem.* 38:191–206.

[23] Kubarych, K.J., Milne, C.J., Lin, S., Astinov, V., Miller, R.J.D. (2002). Diffractive optics-based six-wave mixing: Heterodyne detection of the full $\chi^{(5)}$ tensor of liquid CS_2. *J. Chem. Phys.* 116:2016–2042.

[24] Kalpouzos, C., Lotshaw, W.T., McMorrow, D., Kenney-Wallace, G.A. (1987). Femtosecond laser-induced Kerr responses in liquid CS_2. *J. Phys. Chem.* 91:2028–2030; Ruhman, S., Williams, L.R., Joly, A.G., Kohler, B.,

Nelson, K.A. (1987). Nonrelaxational inertial motion in CS_2 liquid observed by femtosecond time-resolved impulsive stimulated scattering. *J. Phys. Chem.* 91:2237–2240.

[25] Loring, R.F., Mukamel, S. (1985). Selectivity in coherent transient Raman measurements of vibrational dephasing in liquids. *J. Chem. Phys.* 83:2116–2128.

[26] Laubereau, A., Kaiser, W. (1978). Vibrational dynamics of liquids and solids investigated by picosecond light-pulses. *Rev. Mod. Phys.* 50:607–665; Schroeder, J., Schiemann, V.H., Jonas, J. (1977). Raman study of temperature and pressure effects on vibrational relaxation in liquid $CHCl_3$ and $CDCl_3$. *Mol. Phys.* 34:1501–1521; Campbell, J.H., Fisher, J.F., Jonas, J. (1974). Density and temperature effects on molecular-reorientation and vibrational-relaxation in liquid methyl-iodide. *J. Chem. Phys.* 61:346–360; George, S.M., Harris, A.L., Berg, M., Harris, C.B. (1984). Picosecond studies of the temperature-dependent of homogeneous and inhomogeneous vibrational linewidth broadening in liquid acetonitrile. *J. Chem. Phys.* 80:83–94; Fischer, S.F., Laubereau, A. (1975). Dephasing processes of molecular vibrations in liquids. *Chem. Phys. Lett.* 35:6–12; Oxtoby, D.W. (1983). Vibrational-relaxation in liquids-quantum states in a classical bath. *J. Phys. Chem.* 87:30283–30333; Grimbert, D., Mukamel, S. (1981). Nonperturbative approach to collisional dephasing – application to vibrational dephasing in liquids. *J. Chem. Phys.* 75:1958–1963.

[27] Mukamel, S. (1995). *Principles of Nonlinear Optical Spectroscopy.* Oxford University Press, Oxford.

[28] Okumura, K., Tanimura, Y. (1997). First-, third-, and fifth-order resonant spectroscopy of an anharmonic displaced oscillators system in the condensed phase. *J. Chem. Phys.* 106:2078–2095.

[29] Tokmakoff, A., Lang, M.J., Jordanides, X.J., Fleming, G.R. (1998). The intermolecular interaction mechanisms in liquid CS_2 at 295 and 165 K probed with two-dimensional Raman spectroscopy. *Chem. Phys.* 233: 231–242.

[30] Palese, S., Buontempo, J.T., Schilling, L., Lotshaw, W.T., Tanimura, Y., Mukamel, S., Miller, R.J.D. (1994). Femtosecond 2-dimensional Raman spectroscopy of liquid water. *J. Phys. Chem.* 98:12466–12470.

[31] Okumura, K., Tanimura, Y. (1997). Femtosecond two-dimensional spectroscopy from anharmonic vibrational modes of molecules in the condensed phase. *J. Chem. Phys.* 107:2267–2283.

[32] Okumura, K., Tanimura, Y. (1997). Interplay of inhomogeneity and anharmonicity in 2D Raman spectroscopy of liquids. *Chem. Phys. Lett.* 277:159–166.

[33] Tanimura, Y. (1998). Fifth-order two-dimensional vibrational spectroscopy of a Morse potential system in condensed phases. *Chem. Phys.* 233:217–229.

[34] Stratt, R.M., Cho, M. (1994). The short-time dynamics of solvation. *J. Chem. Phys.* 100:6700–6708; Stratt, R.M. (1995). The instantaneous normal-modes of liquids. *Acc. Chem. Res.* 28:201–207.

[35] Ladanyi, B.M., Klein, S. (1996). Contributions of rotation and translation to polarizability anisotropy and solvation dynamics in acetonitrile. *J. Chem. Phys.* 105:1552–1561.

[36] Keyes, T. (1996). Normal mode theory of two step relaxation in liquids: Polarizability dynamics in CS_2. *J. Chem. Phys.* 104:9349–9356; Keyes, T. (1997). Instantaneous normal mode approach to liquid state dynamics. *J. Phys. Chem.* 101:2921–2930.

[37] Saito, S., Ohmine, I. (1998). Off-resonant fifth-order nonlinear response of water and CS_2: Analysis based on normal modes. *J. Chem. Phys.* 108:240–251; Saito, S., Ohmine, I. (1999) Water dynamics: Fluctuation, relaxation, and chemical reactions in hydrogen bond network rearrangement. *Acc. Chem. Res.* 32:741–749.

[38] Murry, R.L., Fourkas, J.T., Keyes, T. (1998). Nonresonant intermolecular spectroscopy beyond the Placzek approximation. I. Third-order spectroscopy. *J. Chem. Phys.* 109:2814–2825; Murry, R.L., Fourkas, J.T., Keyes, T. (1998). Nonresonant intermolecular spectroscopy beyond the Placzek approximation. II. Fifth-order spectroscopy. *J. Chem. Phys.* 109:7913–7922.

[39] Keyes, T., Fourkas, J.T. (2000). Instantaneous normal mode theory of more complicated correlation functions: Third- and fifth-order optical response. *J. Chem. Phys.* 112:287–293.

[40] Ma, A., Stratt, R.M. (2000). Fifth-order Raman spectrum of an atomic liquid: Simulation and instantaneous-normal-mode calculation. *Phys. Rev. Lett.* 85:1004–1007.

[41] Ma, A., Stratt, R.M. (2002). The molecular origins of the two-dimensional Raman spectrum of an atomic liquid. I. Molecular dynamics simulation. *J. Chem. Phys.* 116:4962–4971.

[42] Ma, A., Stratt, R.M. (2002). The molecular origins of the two-dimensional Raman spectrum of an atomic liquid. II. Instantaneous-normal-mode theory. *J. Chem. Phys.* 116:4972–4984.

[43] Denny, R.A., Reichman, D.R. (2001). Mode coupling theory of the fifth-order Raman spectrum of an atomic liquid. *Phys. Rev. E* 63:065101(1–10).

[44] Denny, R.A., Reichman, D.R. (2002). Molecular hydrodynamic theory of nonresonant Raman spectra in liquids: Fifth-order spectra. *J. Chem. Phys.* 116:1987–1994.

[45] Cao, J., Yang, S., Wu, J. (2002). Calculations of nonlinear spectra of liquid Xe. I. Third-order Raman response. *J. Chem. Phys.* 116: 3739–3759; Cao, J., Yang, S., Wu, J. (2002). Calculations of nonlinear spectra of liquid Xe. II. Fifth-order Raman response. *J. Chem. Phys.*116:3760–3776.

[46] van Zon, R., Schofield, J. (2002). Mode-coupling theory for multiple-point and multiple-time correlation functions. *Phys. Rev. E* 65:011106 (1–17); van Zon, R., Schofield, J. (2002). Multiple-point and multiple-time correlation functions in a hard-sphere fluid. *Phys. Rev. E* 65:011107 (1–12).

[47] Kim, J., Keyes, T. (2002). Generalized Langevin equation approach to higher-order classical response: Second-order-response time-resolved Raman experiment in CS_2. *Phys. Rev. E* 65:061102(1–7).

[48] Ma, A., Stratt, R.M. (2003). Selecting the information content of two-dimensional Raman spectra in liquids. *J. Chem. Phys.* 119:8500–8510.

[49] Jansen, T.I.C., Snijders, J.G., Duppen, K. (2000). The third- and fifth-order nonlinear Raman response of liquid CS_2 calculated using a finite field nonequilibrium molecular dynamics method. *J. Chem. Phys.* 113: 307–311.

[50] Jansen, T.I.C., Snijders, J.G., Duppen, K. (2001). Interaction induced effects in the nonlinear Raman response of liquid CS_2: A finite field non-equilibrium molecular dynamics approach. *J. Chem. Phys.*114:10910–10921.

[51] Boeijenga, N.H., Pugzlys, A., Jansen, T.I.C., Snijders, J.G., Duppen, K. (2002). Liquid xenon as an ideal probe for many-body effects in impulsive Raman scattering. *J. Chem. Phys.* 117:1181–1187.

[52] Jansen, T.I.C, Swart, M., Jensen, L., Van Dujinen, P.T., Snijders, J.G., Duppen, K. (2002). Collision effects in the nonlinear Raman response of liquid carbon disulfide. *J. Chem. Phys.* 116:3277–3285.

[53] Jansen, T.I.C., Snijders, J.G., Duppen, K. (2003). Close collisions in the two-dimensional Raman response of liquid carbon disulfide. *Phys. Rev. B* 67:134206(1–5).

[54] Saito, S., Ohmine, I. (2002). Off-resonant fifth-order response function for two-dimensional Raman spectroscopy of liquids CS_2 and H_2O. *Phys. Rev. Lett.* 88:207401(1–4).

[55] Saito, S., Ohmine, I. (2003). Off-resonant two-dimensional fifth-order Raman spectroscopy of liquid CS_2: Detection of anharmonic dynamics. *J. Chem. Phys.* 119:9073–9087.

[56] DeVane, R., Ridley, C., Space, B., Keyes, T. (2003). A time correlation function theory for the fifth order Raman response function with applications to liquid CS_2. *J. Chem. Phys.* 119:6073–6082.

[57] DeVane, R., Ridley, C., Space, B., Keyes, T. (2004). Tractable theory of nonlinear response and multidimensional nonlinear spectroscopy. *Phys. Rev. E* 70:050101(1–4).

[58] Tominaga, K.,Yoshihara, K. (1995). Fifth order optical response of liquid CS_2 observed by ultrafast nonresonant six-wave mixing. *Phys. Rev. Lett.* 74:3061–3064.

[59] Steffen, T., Duppen, K. (1996). Femtosecond two-dimensional spectroscopy of molecular motion in liquids. *Phys. Rev. Lett.* 76:1224–1227.

[60] Tokmakoff, A., Lang, M.J., Larsen, D.S., Fleming, G.R., Chernyak, V., Mukamel, S. (1997). Two-dimensional Raman spectroscopy of vibrational interactions in liquids. *Phys. Rev. Lett.* 79:2702–2705.

[61] Tominaga, K., Keogh, G.P., Naitoh, Y., Yoshihara, K. (1995). Temporally 2-dimensional Raman-spectroscopy of liquids by 6-wave mixing with ultra-short pulses. *J. Raman Spectrosc.*, 26:495–501.

[62] Tominaga, K., Yoshihara, K. (1996).Temporally two-dimensional femtosecond spectroscopy of binary mixture of CS_2. *J. Chem. Phys.* 104:1159–1162.

[63] Tominaga, K., Yoshihara, K. (1996). Fifth-order nonlinear spectroscopy on the low-frequency modes of liquid CS_2. *J. Chem. Phys.* 104:4419–4426.

[64] Tominaga, K., Yoshihara, K. (2000). Concentration dependence of the fifth-order two-dimensional Raman signal. *J. Chin. Chem. Soc.* 47:631–635.

[65] Steffen, T., Duppen, K. (1997). Time resolved four- and six-wave mixing in liquids. 2. Experiments. *J. Chem. Phys.* 106:3854–3864.

[66] Steffen, T., Meinders, N.A.C.M., Duppen, K. (1998). Microscopic origin of the optical Kerr effect response of CS_2–pentane binary mixtures. *J. Phys. Chem. A* 102:4213–4221.

[67] Tokmakoff, A., Fleming, G.R. (1997). Two-dimensional Raman spectroscopy of the intermolecular modes of liquid CS_2. *J. Chem. Phys.* 106:2569–2582.

[68] Tokmakoff, A., Lang, M.J., Larsen, D.S., Fleming, G.R. (1997). Intrinsic optical heterodyne detection of a two-dimensional fifth order Raman response. *Chem. Phys. Lett.* 272:48–54.

[69] Steffen, T., Duppen, K. (1997). Analysis of nonlinear optical contributions to temporally two-dimensional Raman scattering. *Chem. Phys. Lett.* 273:47–54.

[70] Fork, R.L., Martinez, O.E., Gordon, J.P. (1984). Negative dispersion using pairs of prisms. *Opt. Lett.* 9:150–152.

[71] Armstrong, M.R., Plachta, P., Ponomarev, E.A., Miller, R.J.D. (2001). Versatile 7-fs optical parametric pulse generation and compression by use of adaptive optics. *Opt. Lett.* 26:1152–1154.

[72] Goodno, G.D., Astinov, V., Miller, R.J.D. (1999). Diffractive optics-based heterodyne-detected grating spectroscopy: Application to ultrafast protein dynamics. *J. Phys. Chem. B* 103:603–607; Cowan, M.L., Ogilvie, J.P., Miller, R.J.D. (2004). *Chem. Phys. Lett.* 386:184–189; Brixner, T., Stiopkin, I.V., Fleming, G.R. (2004). Tunable two-dimensional femtosecond spectroscopy. *Opt. Lett.* 29:884–886.

[73] Cowan, M.L., Bruner, B.D., Huse, N., Dwyer, J.R., Chugh, B., Nibbering, E.T.J., Elsaesser, T., Miller, R.J.D. (2005). Ultrafast memory loss and energy redistribution in the hydrogen bond network of liquid H_2O. *Nature* 434:199–202.

[74] Maznev, A.A., Crimmins, T.F., Nelson, K.A. (1998). How to make femtosecond pulses overlap. *Opt. Lett.* 23:1378–1380; Maznev, A.A., Nelson, K.A., Rogers, J.A. (1998). Optical heterodyne detection of laser-induced gratings. *Opt. Lett.* 23:1319–1321; Fecko, C.J., Eaves, J.D., Tokmakoff, A. (2002). Isotropic and anisotropic Raman scattering from molecular liquids measured by spatially masked optical Kerr effect spectroscopy. *J. Chem. Phys.* 117:1139–1154.

[75] Golonzka, O., Demirdoven, N., Khalil, M., Tokmakoff, A. (2000). Separation of cascaded and direct fifth-order Raman signals using phase-sensitive intrinsic heterodyne detection. *J. Chem. Phys.* 113:9893–9896.

[76] Berne, B.J., Pecora, R. (1976). *Dynamic Light Scattering with Applications to Chemistry, Biology, and Physics*. New York: Dover Publications, Inc.

[77] Kubarych, K.J., Milne, C.J., Miller, R.J.D. (2003). Fifth-order two-dimensional Raman spectroscopy: A new direct probe of the liquid state. *Int. Rev. Phys. Chem.* 22:497–532.

[78] Hellwarth, R.W. (1977). 3rd-order optical susceptibilities of liquids and solids. *Prog. Quantum Electron.* 5:1–68.

[79] Murry, R.L., Fourkas, J.T. (1997). Polarization selectivity of nonresonant spectroscopies in isotropic media. *J. Chem. Phys.* 107:9726–9740.

[80] Blank, D.A., Fleming, G.R., Cho, M., Tokmakoff, A. (2001). Fifth order two-dimensional spectroscopy of the intermolecular and vibrational dynamics in liquids. In *Ultrafast Infrared and Raman Spectroscopy*. Fayer M. editor. New York: Marcel Dekker Inc.

[81] Astinov, V., Kubarych, K.J., Milne, C.J., Miller, R.J.D. (2000). Diffractive optics based two-colour six-wave mixing: phase contrast heterodyne detection of the fifth order Raman response of liquids. *Chem. Phys. Lett.* 327:334–342.

[82] Tokmakoff, A. (1996). Orientational correlation functions and polarization selectivity for nonlinear spectroscopy of isotropic media. 1. Third order. *J. Chem. Phys.* 105:1–12.

[83] Tokmakoff, A. (1996). Orientational correlation functions and polarization selectivity for nonlinear spectroscopy of isotropic media. 2. Fifth-order. *J. Chem. Phys.* 105:13–21.

[84] Kubarych, K.J., Milne, C.J., Miller, R.J.D. (2003). Heterodyne detected fifth-order Raman response of liquid CS_2: 'Dutch Cross' polarization. *Chem. Phys. Lett.* 369:635–642.

[85] Blank, D.A., Kaufman, L.J., Fleming, G.R. (2000). Direct fifth-order electronically nonresonant Raman scattering from CS_2 at room temperature. *J. Chem. Phys.* 113:771–778.

[86] Kaufman, L.J., Geo, J.Y., Ziegler, L.D., Fleming, G.R. (2002). Heterodyne-detected fifth-order nonresonant Raman scattering from room temperature CS_2. *Phys. Rev. Lett.*, 88:207402 (1–4).

[87] Kaufman, L.J., Saito, S., Ziegler, L.D., Ohmine, I., Fleming, G.R. (2003). In *Ultrafast Phenomena XIII*. Miller, R.J.D., Murnane, M.M., Scherer, N.F., Weiner, A.M., editors, p. 554. Berlin: Springer.

[88] Li, Y.L., Milne, C.J., Jansen, T.l.C., Miller, R.J.D. (in preparation). These calculations were conducted in collaboration with T.l.C. Jansen using a modified version of the previous code used to calculate the $\chi^{(5)}$ response [49–53]. The primary difference is the incorporation of a more accurate compressibility for CS_2 in the FFMD calculations.

[89] Steffen, T., Fourkas, J.T., Duppen, K. (1996). Time resolved four- and six-wave mixing in liquids. I. Theory. *J. Chem. Phys.* 105:7364–7382.

[90] Steffen, T., Duppen, K. (1998). Population relaxation and non-Markovian frequency fluctuations in third- and fifth-order Raman scattering. *Chem. Phys.* 233:267–285.

[91] Ryu, S., Stratt, R.M. (2004). A case study in the molecular interpretation of the optical Kerr effect spectra: Instantaneous-normal-mode analysis of the OKE spectrum of liquid benzene. *J. Phys. Chem. B* 108:6782–6795.

[92] Kubarych, K.J., Milne, C.J., Lin, S., Miller, R.J.D. (2002). Diffractive optics implementation of time- and frequency-domain heterodyne-detected six-wave mixing. *Appl. Phys. B* 74:S107–S112.

[93] Asplund, M.C., Zanni, M.T., Hochstrasser, R.M. (2000). Two dimensional infrared spectroscopy of peptides by phase-controlled femtosecond vibrational photon echoes. *Proc. Natl Acad. Sci. USA* 97:8219–8224.

[94] Cho, M.H., Vaswani, H.M., Brixner, T., Fleming, G.R. (2005). Exciton analysis in 2D electronic spectroscopy. *J. Phys. Chem. B* 109:10542–10556; Brixner, T., Stenger, J., Vaswani, H.M., Fleming, G.R. (2005). Two-dimensional spectroscopy of electronic couplings in photosynthesis. *Nature* 434:625–628.

[95] DeVane, R., Space, B., Perry, A., Neipert, C., Ridley, C., Keyes, T. (2004). A time correlation function theory of two-dimensional infrared spectroscopy with applications to liquid water. *J. Chem. Phys.* 121:3688–3701; Hayashi, T., Mukamel, S. (2003). Multidimensional infrared signatures of intramolecular hydrogen bonding in malonaldehyde. *J. Phys. Chem. A* 107:9113–9131; Scheurer, C., Piryatinski, A., Mukamel, S. (2001). Signatures of beta-peptide unfolding in two-dimensional vibrational echo spectroscopy: A simulation study. *J. Am. Chem. Soc.* 123: 3114–3124.

[96] Saito, S., Ohmine, I. (1997). Third order nonlinear response of liquid water. *J. Chem. Phys.* 106:4889–4893.

[97] Ohmine, I. (1995). Liquid water dynamics: collective motions, fluctuation, and relaxation. *J. Phys. Chem.* 99:6767–6776.

[98] Hura, G., Russo, D., Glaeser, R.M., Head-Gordon, T., Krack, M., Parrinello, M. (2003). Water structure as a function of temperature from X-ray scattering experiments and ab initio molecular dynamics. *Phys. Chem. Chem. Phys.* 5:1981–1991.

[99] Saito, S., Ohmine, I. (1995). Translational and orientational dynamics of a water cluster $(H_2O)108$ and liquid water: Analysis of neutron scattering and depolarized light scattering. *J. Chem. Phys.* 102:3566–3579.

Chapter 2

OPTICAL KERR EFFECT EXPERIMENTS ON COMPLEX LIQUIDS

A Direct Access to Fast Dynamic Processes

Paolo Bartolini, Andrea Taschin, Roberto Eramo, and Renato Torre

Abstract   The time-resolved spectroscopy based on polarization effects represents one of the most sensitive techniques for studying dynamical phenomena in condensed matter. The optical Kerr effect performed with ultra-short laser pulses enables a unique investigation of dynamic processes covering a wide time range, typically from few femtoseconds up to many nanoseconds. This spectroscopic tool is particularly well suited for the measurement of relaxation patterns in complex liquids where several dynamic phenomena, taking place on different time scales, are present. In this chapter we introduce the optical Kerr effect principles, the experimental procedure, and some results from measurements in a number of different complex liquids.

2.1   Introduction

Since 1875, thanks to Kerr's discovery [1], it is known that a static electric field can induce a modification of the optical properties of a liquid. Many years later researchers found out that also an optical electromagnetic field was capable of producing a measurable modification of the dielectric properties, inducing a birefringence effect: the first experimental observation of the optical Kerr effect (OKE) was reported in 1963 [2]. After few years, with the introduction of the first pulsed lasers, spectroscopists discovered the chance to induce in a material a transient birefringence and to measure its relaxation toward the equilibrium [3]. They also realized that this could be a relevant new spectroscopic tool able to collect new information on the dynamical processes present in the material. The spectroscopic research, worked out in the following years, confirmed this forecast beyond the expectations. Two important experimental improvements of this spectroscopic technique have been made. On one hand, the pulsed laser sources have become able to produce very short pulses of high energy in a very reliable way, on the other hand the introduction of the optical heterodyne detection has improved substantially the quality of the OKE data in terms of signal/noise ratio [4] and control of spurious birefringence effects in the signal [5]. During the 80s the first OKE experiments utilizing subpicosecond laser pulses with heterodyne detection were presented [6, 7]. These pioneering works defined the real possibility of the optical Heterodyne Detected Optical Kerr Effect (HD-OKE) in investigating the fast dynamics of simple molecular liquids, and it opened the way to a series of new experimental HD-OKE studies [8–21].

Recently, the OKE technique has been utilized to investigate the dynamics of more complex liquids, like molecular liquids approaching a phase transition [22], liquid crystals [23] and glass-formers [24]. To perform a correct and complete spectroscopic study of these complex systems, the OKE needs to be performed with heterodyne detection [25, 26] over a very large time window; these requirements forced researchers to develop new experimental procedures [26–30]. From these studies clearly arises the need for a more detailed understanding of the general problem of the interaction of the laser pulses with the molecular liquids and the following induced effects, often common to other time-resolved techniques [31–35], see also Chap. 3. All these time-resolved spectroscopic investigations made clear that the study of relaxation phenomena in the time domain gives valuable information, which is complementary to the frequency domain measurements. This is particularly valuable when complex decay patterns and a variety of relaxation channels are present. In some respects, the development of new time-resolved spectroscopies gives also a renew relevance to the HD-OKE experiments that are still able to produce very interesting results on well-studied molecular liquids [36, 37]. In the first Section, we introduce the basic principle of OKE experiments defining the measured signal. In Sect. 2.3, we analyze the OKE response function and its connection with some theoretical models of liquid dynamics. In Sect. 2.4, we present the experimental procedure and set-up implemented by the authors. In the last Section, we summarize some experimental results on simple and complex liquids.

2.2   Time-Resolved OKE Experiments

When a laser pulse passes through a material it produces a local non equilibrium state that induces a modification of the optical properties. This is a transient effect that relaxes back to the equilibrium state through a variety of processes. In a typical pump-probe experiment, a second laser pulse is sent on the material probing the optical modifications induced by the pump pulse. Since the second laser pulse arrives with a controlled delay, it monitors and measures the transient optical excitation and hence the relaxation of the nonequilibrium state. In the time-resolved OKE both the pulses, pump and probe, are linearly

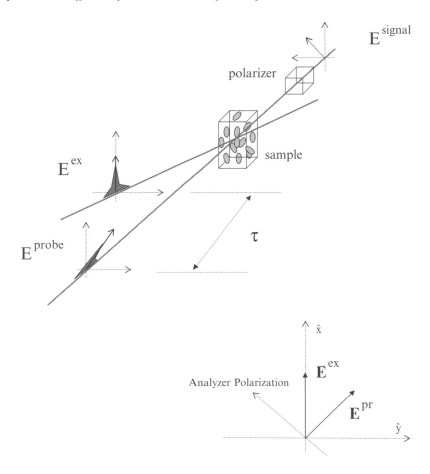

Fig. 2.1 The laser pulse and sample configuration used in a typical OKE experiment

polarized and nonresonant with any electronic state. The pump pulse induces in an isotropic medium (e.g. a liquid or glass) a transient optical anisotropy, as it modifies the index of refraction along the spatial axis parallel to the pump polarization direction, see Fig. 2.1. When no absorption of the laser field is present, just the real part of the refractive index is modified and this effect is called induced birefringence. On the other hand, if a laser absorption is present, also the imaginary part of the refractive index is modified and the effect is called induced dichroism [5]. In this chapter, we will focus only on transparent complex liquids, i.e. electronically nonresonant liquids. Indeed, for completeness, we must note that recently the presence of a very weak dichroic contribution in the HD-OKE signal has been reported also in a medium without any electronic resonance [38] and, moreover, it has been reported that also other type of resonances, as overtones and/or combinations of vibrational bands, could induce a birefringence in the sample [31]. Nevertheless, we will assume in the following that these contributions are negligible in a properly optimized OKE experiment.

2.2.1   A Simple Model

The physical processes describing the field–matter interaction in an OKE experiment are quite complicated and they are defined properly by the nonlinear optics equations [39–41]. In the next chapter, we will outline the rigorous description of the experiment based on the four-wave mixing phenomena equations. Here we introduce a relatively simple model of the OKE experiment, either continuous or time-resolved, that retains all the principal features. This model is based on few intuitive starting approximations that will be proved later. The first basic approximation is the separation between excitation and the probing processes.

2.2.1.1   The Excitation Process

The first laser pulse is linearly polarized in the $\hat{x}$ direction, see Fig. 2.1, and it can be described during its propagation in the sample, according to the following equation, see also the appendix:

$$\mathbf{E}^{\mathrm{ex}}(z,t) = \hat{\mathrm{e}}_x \mathcal{E}^{\mathrm{ex}}(z,t)\, \mathrm{e}^{\mathrm{i}\omega(\frac{\eta_0}{c}z - t)}, \tag{2.1}$$

where η_0 is the equilibrium isotropic index of refraction of the sample. This pulse propagates in the material producing a transient linear birefringence. The laser pulse modifies the real part of the index of refraction, introducing a small anisotropic variation. As a first approximation, we can write the new excited-state refractive index as follows:

$$\eta_{ij}(\tau) = \eta_0 + \Delta\eta_{ij}(\tau), \tag{2.2}$$

$$\Delta\eta_{ij}(\tau) = \left(\begin{array}{cc} \Delta\eta_{xx}(\tau) & 0 \\ 0 & \Delta\eta_{yy}(\tau) \end{array} \right), \tag{2.3}$$

where $\Delta\eta_{ij}$ is the transient birefringence induced by the laser excitation pulse and the i and j indicate the direction of polarization and τ is the delay time between the excitation and probing times. We suppose here non overlapping pulses (i.e. $\tau > 0$) and a medium response time longer than the pulse duration. In the present experimental configuration, because of the initial isotropy of the medium, only the xx (parallel to the excitation) and the yy (perpendicular to the excitation) matrix elements are modified by the excitation beam. The induced anisotropy is defined by the interaction between the excitation pulse and the material; if this is nonresonant the following equation holds:

$$\Delta\eta_{ij}(\tau) \propto \int \mathrm{d}t' \mathcal{R}_{ijxx}(\tau - t') I_{\mathrm{ex}}(t'), \tag{2.4}$$

where $I_{\mathrm{ex}} \propto |\mathcal{E}_x^{\mathrm{ex}}(z,t)|^2$ is the intensity of the laser pulse and $\mathcal{R}_{ijxx}(t)$ is the material response function that defines how the optical properties of the material are modified by the excitation laser pulse (that in the present case has the $\hat{\mathbf{e}}_x$ linear polarization, see (2.1)). Furthermore, the response function defines which modes are excited in the material, by the field–matter interaction, and how these modes evolve in time according to the material dynamics. This function is indeed the more relevant observable to be measured in a time-resolved spectroscopic experiment. In the following chapters, we will analyze the response function in detail as well how this is connected with molecular dynamics present in the materials.

2.2.1.2   The Probing Process

After the sample has been excited, the probe beam propagates in the sample probing the modified index of refraction. We assume the probe beam linearly polarized at 45° with respect to the excitation beam in the $\hat{x}\hat{y}$ plane. The expression of the laser beam before the sample has been reported in the appendix. After the beam has propagated through the sample, the anisotropic index of refraction produces a modification of phase, which is different for the probe components along $\hat{x}$ and $\hat{y}$. Neglecting the effect of the induced birefringence (see appendix) on the probe envelope functions, $\mathcal{E}$, the expression for the probe electric field after the sample is

$$\mathbf{E}^{\mathrm{pr}}(z,t) = \hat{\mathbf{e}}_x E_x^{\mathrm{pr}}(z,t) + \hat{\mathbf{e}}_y E_y^{\mathrm{pr}}(z,t),$$

$$E_x^{\mathrm{pr}}(z,t) = \mathcal{E}^{\mathrm{pr}}(z,t)\, \mathrm{e}^{\mathrm{i}\phi_{xx}(\tau)}\, \mathrm{e}^{\mathrm{i}\omega(\frac{z}{c}-t)},$$

$$E_y^{\mathrm{pr}}(z,t) = \mathcal{E}^{\mathrm{pr}}(z,t)\, \mathrm{e}^{\mathrm{i}\phi_{yy}(\tau)}\, \mathrm{e}^{\mathrm{i}\omega(\frac{z}{c}-t)}, \tag{2.5}$$

where the phase modification produced by the sample, $\phi(\tau)$, according to Eq. (2.3), can be written as:

$$\phi_{xx}(\tau) = \frac{\omega l}{c}\eta_{xx}(\tau) = \frac{\omega l}{c}(\Delta\eta_{xx}(\tau) + \eta_0),$$

$$\phi_{yy}(\tau) = \frac{\omega l}{c}\eta_{yy}(\tau) = \frac{\omega l}{c}(\Delta\eta_{yy}(\tau) + \eta_0), \tag{2.6}$$

where l is the sample length in the $\hat{z}$ direction. The previous equations, (2.5) and (2.6), describe the beam after the propagation through the excited sample. Successively, the probe beam is analyzed by a polarizer crossed with the input probe beam polarization, i.e. $-45°$ with respect to the excitation beam. The beam exciting the crossed polarizer represents the signal field measured in an OKE experiment, $\mathbf{E}^{\mathrm{sg}}(z,t)$. The effect of the analysis polarizer is to project the components of the probe beam according to the following expression [1]:

$$
\begin{aligned}
\mathbf{E}^{\text{sg}}(z,t) &= (\hat{e}_x - \hat{e}_y)[E_x^{\text{pr}}(z,t) - E_y^{\text{pr}}(z,t)] \\
&= (\hat{e}_x - \hat{e}_y)[e^{i\phi_{xx}(\tau)} - e^{i\phi_{yy}(\tau)}]\mathcal{E}^{\text{pr}}(z,t)\, e^{i\omega(\frac{z}{c}-t)} \\
&= (\hat{e}_x - \hat{e}_y)\, e^{i\frac{\omega l}{c}\eta_0}[e^{i\frac{\omega l}{c}\Delta\eta_{xx}(\tau)} - e^{i\frac{\omega l}{c}\Delta\eta_{yy}(\tau)}]\mathcal{E}^{\text{pr}}(z,t)\, e^{i\omega(\frac{z}{c}-t)},
\end{aligned}
\tag{2.7}
$$

where we used (2.6). Considering that the induced variation $\Delta\eta(t)$ is a small effect and showing explicitly only the time-dependent parameters, we can approximate the previous equation as follows:

$$
\begin{aligned}
E^{\text{sg}}(z,t) &\propto i\,[\Delta\eta_{xx}(\tau) - \Delta\eta_{yy}(t)]\,\mathcal{E}^{\text{pr}}(z,t)\, e^{i\omega(\frac{z}{c}-t)} \\
&\propto i\,[\int dt'\,\mathcal{R}_{\text{oke}}(\tau-t')\mathcal{I}(t')]\,\mathcal{E}^{\text{pr}}(z,t)\, e^{i\omega(\frac{z}{c}-t)},
\end{aligned}
\tag{2.8}
$$

where $\mathcal{R}_{\text{oke}} = \mathcal{R}_{xxxx} - \mathcal{R}_{yyxx}$ and $\mathcal{I}(t) \propto [\mathcal{E}(t)]^2$, begin $\mathcal{E}(t)$ the envelope describing the main laser pulse, see appendix. Equation (2.8) defines the signal beam measured in an OKE experiment and it shows that the dynamical information of the material are obtained through the measured response function. In the present derivation we supposed well separated pulses, neglecting the instantaneous electronic contribute. This can be taken into account for a sufficiently thin medium in order to neglect pulse spreading and slipping effects: the final result is that, in Eq. 2.8, $\mathcal{R}_{\text{oke}}(\tau-t')$ should be replaced by $\mathcal{R}_{\text{oke}}(t-t')$.

The signal electromagnetic wave can be detected using two different experimental schemes: the homodyne or the heterodyne detection. In the first detection scheme, the signal beam is sent to a square law detector, typically a photomultiplier or a photodiode followed by a current integrator. In this case, the measured signal can be written according to the following expression:

$$
S_{\text{homo}}(\tau) \propto \int |E^{\text{sg}}(t)|^2\, dt \propto \int dt\mathcal{I}(t-\tau)[\int dt'\mathcal{R}_{\text{oke}}(t-t')\mathcal{I}(t')]^2,
\tag{2.9}
$$

In the heterodyne detection scheme, that will be described in detail in Sect. 2.4.1, the signal field is superimposed with a local field so that the measured signal is linearly connected with the material response. In this case, see also (2.62), the signal can be written as

$$
S_{\text{hete}}(\tau) \propto \int dt\mathcal{I}(t-\tau)\int dt'\mathcal{R}_{\text{oke}}(t-t')\mathcal{I}(t').
\tag{2.10}
$$

If the pulse durations are short compared to the timescales of material dynamics, the intensity can be approximated as $\mathcal{I}(t) \simeq \mathbb{I}\delta(t)$ and so the OKE experiments measure directly in the time domain the response function:

$$
S_{\text{homo}}(\tau) \propto [\mathcal{R}_{\text{oke}}(\tau)]^2\mathbb{I}_{\text{pr}}\mathbb{I}_{\text{ex}}^2,
\tag{2.11}
$$

$$
S_{\text{hete}}(\tau) \propto \mathcal{R}_{\text{oke}}(\tau)\mathbb{I}_{\text{pr}}\mathbb{I}_{\text{ex}}.
\tag{2.12}
$$

We would like to stress how the time resolution in this OKE experiment is determined by the laser pulse duration. Indeed, other experimental parameters are important, as we will see in Sect. 2.4, e.g. the optical delay line that defines the delay τ. So if short laser pulses are available, fast dynamic phenomena can be directly revealed and measured in the time domain.

2.2.2 The Nonlinear Optics Framework

The basic equations describing the OKE can be introduced within the framework of nonlinear optics and spectroscopy [39–41].

In this approach, the interaction of an electromagnetic field, E, with the matter is described on the basis of a series expansion of field power. So basically the polarization induced in a medium is the sum of separated effects that are linear, quadratic, cubic, and so on, in E. Each of these effects shows a variety of phenomena that can be isolated and described by the nonlinear optics framework. The optical Kerr effect is a third-order nonlinear process (i.e. it is produced by the cubic term in the polarization expansion); more specifically it is a four-wave mixing process. Two waves are associated with the excitation laser, thus acting twice in the field expansion, and a third wave with the probe beam. These three electromagnetic waves interact in the material producing a third-order polarization, via the third-order susceptibility. The nonlinear polarization is source of a new electromagnetic field that represents the fourth wave and the signal from the spectroscopic point of view, differing by the probe beam only for the polarization state.

The third-order polarization can be written as [40]

$$P_i^{(3)}(t) = \sum_{jkl} \int \mathrm{d}t_1 \int \mathrm{d}t_2 \int \mathrm{d}t_3$$
$$\mathcal{R}_{ijkl}^{(3)}(t - t_1, t - t_2, t - t_3) E_j(t_1) E_k(t_2) E_l(t_3), \quad (2.13)$$

where $P_i^{(3)}$ is the ith component of the third-order polarization, $\mathcal{R}^{(3)}$ is the general third-order response tensor of the material. In (2.13) we assumed that the macroscopic nonlinear polarization density, $P^{(3)}(\mathbf{r}, t)$, depends locally only on the field, $E(\mathbf{r}, t)$, at its own position $\mathbf{r}$. With this assumption we have omitted the reference to $\mathbf{r}$ in the response tensor $\mathcal{R}^{(3)}$ [41].

The $\mathcal{R}^{(3)}$ tensor is a very complex function that defines the coupling of the optical electromagnetic fields with the material and how these are modified by the structural and dynamical properties of the medium. In the general definition of such tensor many nonlinear optical phenomena are included. A full description of this physical property is a tremendous task even at the simpler level. Nevertheless, a few general properties and some approximations help in partially reducing the $\mathcal{R}^{(3)}$ complexity.

As previously introduced, we are focusing on the nonresonant OKE where all the laser pulses, excitation, and probe are characterized by the same frequency not resonant with any material electronic state. In this case the Born-Oppenheimer (B-O) approximation is valid and a simplified expression for the third-order response can be introduced [41]:

$$
\begin{aligned}
\mathcal{R}^{(3)}_{ijkl}(t - t_1, t - t_2, t - t_3) &= \gamma^{(e)}_{ijkl}\delta(t - t_1)\delta(t - t_2)\delta(t - t_3) \\
&+ \mathcal{R}^{(\mathrm{nucl})}_{ijkl}(t - t_2)\delta(t - t_1)\delta(t_2 - t_3). \quad (2.14)
\end{aligned}
$$

The third-order response can be divided in an instantaneous response due to the electronic hyperpolarizability plus a second noninstantaneous response defined by the nuclear dynamics and hence by the molecular dynamics. This second part includes all the relevant dynamic information in a complex liquid. The proper definition of the nuclear response function will be introduced later, here we recall only the symmetry rules that must be verified by it. In an isotropic medium, as a liquid or a glass, $\mathcal{R}^{(\mathrm{nucl})}$ is subject to several symmetry rules [41]. According to our experimental geometry, there are 8 nonvanishing tensor elements, but only 2 independent components:

$$
\mathcal{R}^{(\mathrm{nucl})}_{xxxx} = \mathcal{R}^{(\mathrm{nucl})}_{yyyy} \ , \ \mathcal{R}^{(\mathrm{nucl})}_{xxyy} = \mathcal{R}^{(\mathrm{nucl})}_{yyxx}
$$

$$
\mathcal{R}^{(\mathrm{nucl})}_{xyxy} = \mathcal{R}^{(\mathrm{nucl})}_{yxyx} = \mathcal{R}^{(\mathrm{nucl})}_{yxxy} = \mathcal{R}^{(\mathrm{nucl})}_{xyyx} = \frac{1}{2}(\mathcal{R}^{(\mathrm{nucl})}_{xxxx} - \mathcal{R}^{(\mathrm{nucl})}_{yyxx}). \quad (2.15)
$$

The general definition of third-order polarization (2.13) is simplified by the B-O approximation, hence using the response definition reported in (2.14) we get [41]

$$
\begin{aligned}
P^{(3)}_i(t) &= \gamma^{(e)}_{ijkl}E_j(t)E_k(t)E_l(t) \\
&+ E_j(t)\int dt' \mathcal{R}^{(\mathrm{nucl})}_{ijkl}(t - t')E_k(t')E_l(t'). \quad (2.16)
\end{aligned}
$$

This expression is still quite a general definition that can be further simplified in order to describe the OKE experiments.

In the present OKE experiment, the electromagnetic fields are characterized by the same frequency and wave-vector with different polarization states. Furthermore, we can assume the B-O separation as reported in (2.16). So, neglecting the instantaneous electronic response, considering the time separation between excitation and probe pulses, and verifying the phase matching conditions and the energy conservation, the induced polarization becomes [40]

$$
\begin{aligned}
P^{(3)}_i(t) &= E^{\mathrm{pr}}_j(t)\Delta\chi_{ij}(t), \\
\Delta\chi_{ij}(t) &= \int dt' \mathcal{R}^{(\mathrm{nucl})}_{ijxx}(t - t')E^{\mathrm{ex}}_x(t')[E^{\mathrm{ex}}_x(t')]^*, \quad (2.17)
\end{aligned}
$$

where we assume a sum over repeated indexes. This equation proves that a separation between the excitation process and the probing process, see Sect. 2.2.1, is correct. Furthermore, the second equation is equivalent to (2.4) reported in the previous section.[2] In a general case, when resonant and nonresonant laser interactions are present, $P^{(3)}(t)$ has a different phase from the probe beam. The part of $P^{(3)}(t)$ that is in-phase defines the birefringence, whereas the part that is $\pi/2$ out-of-phase describes the dichroic effect. This separation of the two polarization contributions together with heterodyne detection (see Sect. 2.4.1) allows disentanglement of the birefringence from the dichroic signal: when the local oscillator is $\pi/2$ out-of-phase with the probe, only the birefringence signal is detected; if the local oscillator is in phase only the dichroic signal is measured [7].

Introducing now the probe and excitation field polarizations, according to the experimental configuration shown in Fig. 2.1, and the symmetry rules on the nonlinear response valid for an isotropic medium, see Eq. (2.15), the induced polarization components are

$$\mathbf{P}^{(3)}(t) = \hat{e}_x P_x^{(3)}(t) + \hat{e}_y P_y^{(3)}(t),$$

$$P_x^{(3)}(t) = E_x^{\text{pr}}(t') \int dt' \mathcal{R}_{xxxx}(t - t')|E_x^{\text{ex}}(t')|^2,$$

$$P_y^{(3)}(t) = E_y^{\text{pr}}(t') \int dt' \mathcal{R}_{yyxx}(t - t')|E_x^{\text{ex}}(t')|^2, \tag{2.18}$$

where we have omitted the (nucl) superscript in order to simplify the notation.

As already mentioned, the third-order polarization induced in the material produces the fourth field: $\mathbf{E}^{\text{4th}}$. This field is defined by the differential equation that defines the wave propagation in the medium, where $P^{(3)}$ is the source term [40]. According to an approximated solution of this equation we found that $\mathbf{E}^{\text{4th}} \propto i\mathbf{P}^{(3)}$. In other words, the 4th electromagnetic field is the probe beam modified by the excited material and it corresponds to (2.5) previously derived. Following the present experimental configuration, the polarization state of this field is analyzed by a crossed polarizer, see Fig. 2.1. The output field is the measured signal field, $\mathbf{E}^{\text{sg}}$. This optical device produces a projection of the $\mathbf{E}^{\text{4th}}$ polarization components, so that the following signal field results:

$$\mathbf{E}^{\text{sg}}(t) \propto i(\hat{e}_x - \hat{e}_y)E^{\text{pr}}(t) \int dt' [\mathcal{R}_{xxxx}(t - t') - \mathcal{R}_{yyxx}(t - t')]|E_x^{\text{ex}}(t')|^2, \tag{2.19}$$

where the E^{pr} is the original probe beam (i.e., before it passes through the material). Hence, considering the following definition, derived from the symmetry rules

$$\mathcal{R}_{\text{oke}}(t) \propto \mathcal{R}_{xyxy}(t) = \frac{1}{2}[\mathcal{R}_{xxxx}(t) - \mathcal{R}_{yyxx}(t)], \tag{2.20}$$

and the probe and excitation laser pulses expressions, see (2.A.4) in the appendix, the previous equation can be further simplified to

$$\mathbf{E}^{sg}(t) \propto i\,(\hat{e}_x - \hat{e}_y)\,\mathcal{E}^{pr}(t)\,e^{i\omega(\frac{z}{c}-t)}\int dt'\,\mathcal{R}_{oke}(t-t')[\mathcal{E}_x^{ex}(t')]^2. \quad (2.21)$$

This equation shows how the signal field, according to the introduced approximations, is a plane wave with a slowly varying envelope, see appendix, and it is equivalent to (2.8).

So we found that the nonlinear optics model provides a rigorous description of OKE experiment, where all the relevant approximations are physically defined. We would like to stress again, that the B-O approximation is indeed fundamental. We also showed that this nonlinear optics model explains the basic hypothesis introduced in the simple model, previously described.

2.3   The Response Function

All the material information measurable in a OKE experiment are contained in the third-order response, $\mathcal{R}^{(3)}$, introduced in the previous Sect. 2.2.2. When a nonresonant experiment is performed the Born-Oppenheimer approximation applies and the relevant function becomes the material response, $\mathcal{R}_{ijkl}$, see (2.14) and (2.17). In the B-O model, the response function can always be directly connected with the time-dependent correlation function of the quantum linear susceptibility, χ_{ij}, [41]:

$$\mathcal{R}_{ijkl}(q,t) \propto \frac{i}{\hbar}\,\theta(t)\,\langle [\chi_{ij}(q,t)\,,\,\chi_{kl}(-q,0)]\rangle, \quad (2.22)$$

where $[\cdot]$ is the commutator, the $\langle\cdot\rangle$ is the ensemble average, and $\theta(t)$ the Heaviside step function. This expression in the classical limit becomes

$$\mathcal{R}_{ijkl}(q,t) \propto -\frac{\theta(t)}{k_B T}\frac{\partial}{\partial t}\,\langle \chi_{ij}(q,t)\,\chi_{kl}(-q,0)\rangle, \quad (2.23)$$

where k_B is the Boltzmann constant and T is the temperature, and we replaced the quantum linear susceptibility with the classical susceptibility tensor. For the sake of simplicity, we wrote the full susceptibility functions in the previous expressions, but we should recall that only the time dependent or fluctuating part, $\delta\chi_{ij}$, is relevant for the response functions. Often in the literature the susceptibility is replaced by the full dielectric tensor since their fluctuating part coincide: $\delta\varepsilon_{ij} = 4\pi\delta\chi_{ij}$, in the CGS units.

Indeed the B-O approximation produces a relevant simplification of the $\mathcal{R}^{(3)}$ tensor. It allows a separation of the electronic and nuclear responses (see (2.14)) and to express the latter in a useful way (2.23). Nevertheless, the $\chi_{ij}(q,t)$ tensor remains a very complex physical variable that can be detailed only using severe approximations, as we will see later.

In the previous equations, we reported the general response function depending also on the wave-vector q; this dependence was neglected in all the previous expressions since this aspect is not relevant in the OKE experiments. In fact in any scattering experiment, linear or nonlinear, the measured signal is defined through the space integration over the sample volume. This integration turns out to be a Fourier-transformation of the response function from the real space, r coordinates, to the reciprocal space, q wave-vectors. So $\mathcal{R}^{(3)}_{ijkl}$ is a time-q dependent function, where the q value is defined by the experimental configuration. As will be detailed in the next chapter, the phase matching condition defines the relation between the angles of the interacting electromagnetic fields, in particular the angle between the two excitation fields, defining the q vector value. In the OKE technique, they are indeed the same field acting twice, so this angle is zero and consequently the wave-vector is zero. So only the response function characterized by $q \sim 0$ is active in the OKE experiments. This is the reason why we do not report the q-dependence in the equations introduced in the previous paragraphs. A different situation is present in the light scattering (LS) experiments, see [42], or in the transient grating experiment, see Chap. 3. In the former, q is defined by the scattering angle. In the latter it is defined, as in the OKE, by the angle formed by the two excitation fields. In a transient grating, these have different directions of propagation, so the response function is characterized by a nonzero wave-vector.

Expression (2.23) of the response function shows that this experiment verifies the linear response theorem [43]. In fact, when the excitation fields are not resonant and relatively weak, the response is defined by the equilibrium fluctuations, hence the response features are not dependent on the excitation fields' intensities. The validity of this linear approximation can be directly checked in the experiments by measuring the response pattern at different excitation intensities and verifying that it does not change.

As we showed, cf. (2.23) and (2.20), the dynamic observable in an OKE experiment, and also in the forward depolarized light scattering (see also Sect. 2.4.2) is the correlation of the susceptibility or dielectric fluctuations:

$$\Phi^{\text{oke}}_{\chi\chi}(t) = \langle \chi_{xy}(t)\,\chi_{xy}(0)\rangle. \qquad (2.24)$$

The optically accessible dynamical information on the material is contained in the susceptibility tensor, $\chi_{ij}(q,t)$, and the correlation of its fluctuation. This tensor is a very complex material property. We have to deal with two problems. The first one concerns the proper definition of the tensor on the basis of the fundamental physical parameters of the material. The second challenge is the construction of a theoretical model able to describe the dynamics of such physical parameters. Both are typical many-body problems that can be undertaken only using strong approximations. Here we just want to introduce some basic models, useful to describe the susceptibility tensor dynamics in complex liquids [42, 44, 45].

2.3.1   The Linear Susceptibility Tensor and Its Correlation

The basic physical models of the susceptibility tensor,[3] which are indispensable to connect the response function with the dynamics of simple and complex liquids, will be described here.

The susceptibility tensor can be separated into two main contributions [42, 44]:

$$\chi_{ij}(q,t) = \chi_{ij}^{\mathrm{IM}}(q,t) + \chi_{ij}^{\mathrm{II}}(q,t), \tag{2.25}$$

χ^{IM} describes the Isolated-Molecule term, the χ_{ij}^{II} represents the Interaction-Induced. The first term is defined on the basis of the polarizability of the individual molecule,[4] the second term describes the modification of the χ tensor generated by the intermolecular effects. This interaction can be described at the simplest approximation by the dipole-induced-dipole phenomena. The contributions to the tensor of the II effects, and hence to the measured signals, are typically weak in the liquids formed by molecules with anisotropic polarizability, even if their effects can not be neglected a priori [42, 44]. Indeed, the II is a complex phenomena and in the molecular liquids it can be calculated only numerically by computer simulation [46, 47]. To introduce a simple view we will not explicitly consider the II in this chapter. So in the following we will approximate

$$\chi_{ij}(q,t) \simeq \chi_{ij}^{\mathrm{IM}}(q,t). \tag{2.26}$$

The χ^{IM} tensor can be described on the basis of different theoretical approaches. Here we introduce the two more relevant ones with the aim of describing the response function in complex liquids. One model uses a microscopic approach and it starts from the molecular parameters. The other applies a coarse-graining procedure introducing a few mesoscopic or hydrodynamic variables.

2.3.1.1   The Microscopic Variables

The susceptibility tensor, according to the IM approximation, can be directly connected with the molecular parameters in the reciprocal space [42, 44]:

$$\chi_{ij}(q,t) \simeq \sum_n \alpha_{ij}^n(t)\, e^{iq \cdot r_n(t)}, \tag{2.27}$$

where α^n is the polarizability of the individual molecule in the r_n position and the n index runs over the macroscopic ensemble of molecules. According to this expression, the interaction of the electric field with the liquid is defined simply as the sum of single molecule scattering processes. The time dependence of the χ tensor, through the α and r variables, is ruled by the molecular dynamics.

In this equation, there are no hypotheses about the dynamics of the molecules, which is defined by the complete Hamiltonian of the liquid.

To connect the susceptibility tensor with the molecular dynamics, it is useful to separate the intramolecular from the intermolecular dynamics. This separation is applicable when these two dynamics are decoupled.

The intramolecular dynamics is usually described on the basis of vibrational normal modes, $V_v^n(t)$ where n index individuates the molecule and v the vibrational mode. Limiting the mode expansion to the first order, we can write the molecular polarizability as

$$\alpha_{ij}^n(t) \simeq \tilde{\alpha}_{ij}^n(t) + \sum_{n,v} b_{ij}^{n,v}(t) V_v^n(t), \qquad (2.28)$$

where $\tilde{\alpha}$ indicates the molecular polarizability with the nuclear configuration at the equilibrium (i.e. rigid molecule approximation), and $b_{ij}^{n,v}(t) = \left[\frac{\partial \alpha_{ij}^n(t)}{\partial V_v^n} \right]$. The expression $\tilde{\alpha}$ can be better defined when the molecule has a polarizability characterized by high symmetry. In particular for symmetric-top molecules (i.e. molecules that have a full axis of rotation) the polarizability can be separated in an isotropic tensor, $\alpha \delta_{ij}$, and a symmetric traceless tensor, $\beta Q_{ij}^n(t)$, [42,44,45]:

$$\tilde{\alpha}_{ij}^n(t) \simeq \alpha \delta_{ij} + \beta Q_{ij}^n(t), \qquad (2.29)$$

where α and β are the isotropic and anisotropic part, respectively, of the molecular polarizability. The Q_{ij}^n is the orientational tensor that can be defined as $Q_{ij}^n = (\hat{u}_i \hat{u}_j - \frac{1}{3} \delta_{ij})$, where $\hat{u}$ is the unit vector individuating the molecular axis of symmetry.

In the previous equations, the intramolecular dynamics is defined by the vibrational coordinates, $V_v(t)$, whereas the intermolecular dynamics is contained in several parameters defining the susceptibility tensor. It appears explicitly in the molecule translational coordinates, $r_n(t)$, and implicitly in the rigid-molecule polarizability, $\tilde{\alpha}_{ij}^n$, and mode expansion coefficients, $b_{ij}^{n,v}$. In fact both dependent by the molecule orientational coordinates and hence on the rotational dynamics.

When we introduce the definition of (2.27), (2.28) and (2.29) in the correlation function we get the complete response function measured in any light scattering experiment, apart from the II contributions. This response describes the spectrum from the so called Rayleigh-Brillouin to the Raman roto-vibrational scattering.

In the OKE response only the off-diagonal elements, $i \neq j$, of the dielectric tensor are involved, so the isotropic polarizability, α, does not contribute. Furthermore, the experimental q wave-vector value is zero, hence the translational part of (2.27) is not relevant. Concerning the OKE response the susceptibility tensor definition can be simplified as

$$\chi_{xy}(q,t) \simeq \sum_n \beta Q_{xy}^n(t) + \sum_{n,v} b_{xy}^{n,v}(t) V_v^n(t). \qquad (2.30)$$

Using this dielectric tensor in the correlation function (2.24) and supposing no correlation between the vibrational modes of different molecules, we get the following contributions:

$$\Phi_{\chi\chi}^{\text{oke}}(t) = \Phi^{\text{Rot}}(t) + \Phi^{R-V}(t),$$

$$\Phi^{\text{Rot}}(t) = \sum_{n,m} \beta^2 \left\langle Q_{xy}^n(t)\, Q_{xy}^m(0) \right\rangle, \qquad (2.31)$$

$$\Phi^{R-V}(t) = \sum_{n,v} \left\langle b_{xy}^{n,v}(t) V_v^n(t)\, b_{xy}^{n,v}(0) V_v^n(0) \right\rangle. \qquad (2.32)$$

The correlation separates into two main parts: one is purely determined by the rotational dynamics, Φ^{Rot}, the other one, $\Phi^{R-V}(t)$, by the intramolecular vibration, V_v^n, modulated by the rotational processes via the $b^{n,v}$ coefficients. Both these two contributions are typically present in the OKE signal.

The $R-V$ correlation produces a fast oscillating decay whose relative frequencies and damping rates correspond to the Raman lines [7,9]. This part of the OKE is sometimes called the Raman-induced Kerr effect [40]. From the experimental point of view, the intensity of the $R-V$ is strongly dependent by the laser pulse duration. Typically, short pulses of duration 50 fs or less are able to excite intramolecular vibrations by the Raman-induced Kerr effect since their corresponding bandwidth is large enough to reach the Raman-active vibrations present in the organic molecules.[5] If the pulse is of a longer duration, the effective bandwidth is reduced and the higher frequency dynamics can be cut out so that they do not contribute to the signal [27].

The Rot correlation produces a complex relaxation pattern that has to be addressed to the whole rotational dynamics inclusive of librational and diffusive dynamics. This dynamics has been extensively studied by OKE experiments in simple molecular liquids. For example, the liquid phases of benzene or carbon disulfide are particularly interesting systems. In fact, they can be considered symmetric-top rigid molecules (i.e., the molecular polarizability has a cylindrical symmetry and all the vibrational modes have very high frequencies, not excitable by the laser pulses spectrum). Even in such simple molecular liquids, the interpretation of several dynamics is still open to debate [19, 20, 36] (see also Sect. 2.5.1).

On the basis of 2.31, two different contributions to the correlation can be identified:

$$\Phi^{\text{Rot}}(t) = \Phi^{\text{self}}(t) + \Phi^{\text{cross}}(t),$$

$$\Phi^{\text{self}}(t) = \sum_n \beta^2 \left\langle Q_{xy}^n(t)\, Q_{xy}^n(0) \right\rangle, \qquad (2.33)$$

$$\Phi^{\text{cross}}(t) = \sum_{n\neq m} \beta^2 \left\langle Q_{xy}^n(t)\, Q_{xy}^m(0) \right\rangle, \qquad (2.34)$$

where Φ^{self} describes the orientational self-correlation of the single molecule and Φ^{cross} the cross-correlation between different molecules.

2.3.1.2   The Mesoscopic Variables

In the previous section we connected the susceptibility tensor directly to the microscopic parameters, the molecular polarizabilities. Indeed, this point of view clarifies the molecular aspects of the χ tensor, but the connection with the dynamical model is complicated by the extremely large number of variables that must be considered. An alternative possibility is represented by a coarse-grained definition of the susceptibility function, in fact if we disregard to the microscopic information we gain a more direct link to the dynamic model. The polarizability of a rigid symmetric-top molecule can be described by (2.29), hence the dielectric tensor becomes

$$\chi_{ij}(q,t) \simeq \sum_n \left[\alpha \delta_{ij}\, e^{iq \cdot r_n(t)} + \beta Q_{ij}^n(t)\, e^{iq \cdot r_n(t)} \right]. \qquad (2.35)$$

It is worth noting that the first term on the right hand side of this equation is time-dependent only through the translational molecular degree of freedom, r_n, since the isotropic part of the polarizability is not dependent on the molecule orientation. This term is directly connected with the microscopic numerical density:

$$\rho(q,t) = \sum_n e^{iq \cdot r_n(t)}. \qquad (2.36)$$

Conversely the second term is time-dependent through the molecular rotational degree of freedom since the definition of tensor Q_{ij}^n is in the laboratory coordinate, and so in this term there are both the translational and rotational degrees of freedom. We can define the orientational tensor as

$$Q_{ij}(q,t) = \sum_n Q_{ij}^n(t)\, e^{iq \cdot r_n(t)}. \qquad (2.37)$$

This tensor is clearly determined by the distribution of the molecular axes. Indeed when $q \sim 0$, as in the specific case of OKE response, this tensor is completely defined by the molecular axis distribution.

To introduce the mesoscopic level in these expressions, we must consider a coarse-graining procedure. This is indeed a nontrivial statistical method that allows averaging of the microscopic functions, previously reported, obtaining a mesoscopic function. Following this approach we can define [48, 49]

$$\tilde{\rho}(q,t) = \langle \rho(q,t) \rangle_{c-g}, \qquad (2.38)$$

$$\widetilde{Q_{ij}}(q,t) = \langle Q_{ij}(q,t) \rangle_{c-g}, \qquad (2.39)$$

where with the notation $\langle \rangle_{c-g}$ we indicate the partial averaging defined by the coarse-graining procedure. This average is evidently different from the ensemble average that defines the correlation function. After this coarse-graining average, the microscopic functions become mesoscopic observables where the microscopic information is averaged out. These functions show a smoother space/time dependence and verify the condition of local thermodynamic equilibrium. Hence we can rewrite the susceptibility as

$$\widetilde{\chi}_{ij}(q,t) \simeq a\,\tilde{\rho}(q,t)\,\delta_{ij} + b\,\widetilde{Q}_{ij}(q,t), \qquad (2.40)$$

where a and b are the averaged coefficients derived from the isotropic and anisotropic parts, respectively, of the molecular polarizability [48,49].

Expression (2.40) is a very useful definition of the susceptibility tensor since it allows a direct link with the hydrodynamic models of complex liquids, as we will show in the next section.

If we focus our attention on the OKE response, only the off-diagonal elements, $i \neq j$, of the susceptibility tensor are involved and the q-dependence can be neglected. So we can simply consider

$$\chi_{ij}(t) \simeq b\,\widetilde{Q}_{ij}(t). \qquad (2.41)$$

Hence the relevant correlation function becomes

$$\Phi_{XX}^{\text{oke}}(t) = \Phi_{QQ}(t) = b^2 \left\langle \widetilde{Q_{xy}}(t)\,\widetilde{Q_{xy}}(0) \right\rangle, \qquad (2.42)$$

where only the rotational degrees of freedom are relevant. In contrast to the previous microscopic expressions, (2.31) and (2.32), no intramolecular dynamics is present. Furthermore, since the $\widetilde{Q_{xy}}$ function is mesoscopic, no separation between the *self* and *cross* correlation is applicable.

2.3.2   The Model of Liquid Dynamics

The physics of molecular liquids has been described through many different approaches and models, reported widely in the literature. The general framework of the liquid physics can be found in the following references: [50–53], which includes the more recent developments on complex liquids. The basic hydrodynamic point of view of the liquid phase can be found in: [54,55]. Finally the fundamental spectroscopic techniques devoted to the liquid investigations are reported in [42,56,57].

Nevertheless, the present physical models on liquid phases are far from reaching a comprehensive picture of the dynamical scenario. Focusing on complex liquids, even the single dynamic process has been, very often, only partially elucidated. Here we recall some of the typically exploited models to describe and to address the relaxation processes measured in an optical scattering experiment, both OKE and LS. According to a simplified scheme, these models can be classified into two main groups.

The first group starts from a microscopic description of the liquid by the molecular variables and usually estimates the equilibrium average of only a few variables on the basis of a proper statistical model. This procedure allows the evaluation of some dynamic parameters of the liquid susceptibility, as introduced previously in the microscopic model. Recently this approach deeply benefits from computer simulations that enable the extraction of a valid molecular picture of the liquid dynamics from the OKE signal [58, 59].

The second series of models is based on the mesoscopic definition of the liquid variables, as introduced in the previous paragraph, and they focus on the collective dynamic processes. Typically the mesoscopic variables, or their correlators, follow few equations of motion derived from the basic conservation laws [42, 54, 55] or the projection procedure introduced in the Mori-Zwanzig theory [42, 52]. The solution of these equations determines the liquid dynamics. Recently these types of models have encountered a renewed interest, thanks to their application to visco-elastic and glassy liquids [48, 49, 60–62].

Here we just want to recall the main characteristics of the theoretical models, of both types, as they have been used in order to describe the relaxation phenomena measured by the OKE experiments. We focus on a few interpretations of the OKE relaxation. One, immediately following, belongs to the first group; the others relate to the second category.

2.3.2.1 Relaxation Processes and Local Vibrations

Historically, a large amount of OKE data on relatively simple molecular liquids have been described according to a basic separation between the fast and the slow dynamics [57]. Indeed, this decoupling of the molecular dynamics is not always appropriate, but it has been and is widely used [20]. In this framework the OKE correlator (see (2.24)) can be divided as

$$\Phi^{\text{oke}}_{\chi\chi}(t) = \Phi^{\text{fast}}(t) + \Phi^{\text{slow}}(t). \tag{2.43}$$

The proper definition of these two correlators is the fundamental issue. Very often the slow correlator is addressed to the orientational self-correlation of the single rigid molecule, see (2.33), which according to the *Debye-Stokes-Einstein model* (DSE) can be described as a pure Brownian diffusive process [19, 42]. Hence the slow correlator becomes

$$\Phi^{\text{slow}}(t) \equiv \Phi^{\text{self}}(t) = \Sigma_n A_n \, e^{-\frac{t}{\tau_n}}, \tag{2.44}$$

where the number of exponentials and relaxation times, τ_n, are defined by the molecular symmetry. For a true-symmetric-top molecule (i.e., a molecule having a cylindrical symmetry in the diffusion and the polarizability tensors, with coinciding space axis), the $\Phi^{\text{slow}}(t)$ decays as a single exponential characterized by a relaxation time, called the tumbling orientational time. The relaxation times can be connected to the liquid shear viscosity, η, by the following relation [42]

$$\tau_n = \frac{\eta V_{\text{eff}}}{k_B T} \tag{2.45}$$

where V_{eff} is the effective hydrodynamic volume of the molecule, k_B is the Boltzmann constant, and T is the temperature. The definition of V_{eff} is a crucial point of the DSE model and it takes into account, at a very simple and phenomenological level, some of the intermolecular dynamics. To improve the hydrodynamic definition of the (2.45), the V_{eff} has been connected to the real molecular volume, V_m, by a multiplicative coefficient, f, being $V_{\text{eff}} = fV_m$. The parameter f defines the boundary condition and it can be calculated on the basis of a geometrical definition of the molecule [19].

Moreover, as shown in (2.33) and (2.34) the OKE signal has a collective contribution even if the sample is approximated as an ensemble of rigid symmetric molecules. Hence a more appropriate approximation for the slow OKE signal should take into account the cross-correlation term: $\Phi^{\text{slow}}(t) \equiv \Phi^{\text{self}}(t) + \Phi^{\text{cross}}(t)$. The evaluation of the cross-term is intrinsically much more complicated than the self-term since it is a many-body function. A possible analysis of such term can be undertaken using the Mori-Zwanzig theory [42,56], where the many-body problem can be reduced to a few "primary/slow" dynamic variables averaging the "secondary/fast" variables. According to this theoretical procedure, Keyes and Kivelson [63] showed that the "collective" rotational correlation function, i.e. $\Phi^{\text{slow}}(t)$, can be approximated again as a single exponential decay with a normalized relaxation time: $\tau_c = \frac{g}{j}\tau_n$, where the g and j factors have a precise statistical definition.

The fast correlator (See (2.43)) describes all the liquid dynamics that is not a pure diffusive process. So, on the fast time scale, the liquid is taken to be in "frozen" structures. The molecules can vibrate/librate but their mean positions are substantially fixed. The fast vibrational/librational dynamics is collective but typically does not show correlation over many molecules, and so it can be studied as a local dynamics. There are no exact models able to describe such many-body problems [57], but there exist different phenomenological approaches. They are a natural progression of the long-standing problem of cage structures in molecular liquids [64].

The interpretation of OKE data based on this reported model has been only partially successful. Indeed, the separation between fast and slow dynamics is often arbitrary [36]. The picture of slow/diffusive processes as a single molecule moving in the surrounding molecules, if acceptable for a massive solute molecule in a solvent of smaller molecules, it is clearly oversimplified for neat liquids [19]. In particular, in complex liquids the important role of collective motion and of cooperative effects for the long time dynamics has been demonstrated and the single molecule picture appears as an oversimplified one. Also the generalization introduced by Keyes and Kivelson is often not able to reproduce some of the simpler results obtained from OKE experiments, in particular the nonsingle-exponential decay of the signal present in the slow time scale. Furthermore, the interpretation of OKE signal relaxation in the fast/vibrational time scale remains an open issue, even in the simplest molecular liquids [20, 36, 59].

2.3.2.2   Hydrodynamic Models

Hydrodynamic models have been for a long time utilized to describe the spectrum of light scattering from complex liquids [42], but only recently have they been employed to interpret time-resolved spectroscopic data. As we introduced, this model starts from a mesoscopic definition of the liquid variables. We found that, in order to describe the OKE signal, a single correlator (see 2.42) can be retained. Since the $\widetilde{Q}(t)$ is a coarse-grained variable, its dynamics can be described by the hydrodynamic equations. Indeed only recently has been suggested an hydrodynamic model that takes into account translation/density, $\tilde{\rho}(q,t)$, and rotational variables, $\widetilde{Q_{ij}}(q,t)$, with the proper coupling mechanism [48,49]. In the hydrodynamic model introduced by Pick et al. [49], it is proved that the direct coupling between slow density dynamics and orientation variables can be neglected in the evaluation of OKE response (i.e. in the limit of small q wave-vector). The basic dynamic equation can be written as[6]:

$$\ddot{Q}_{xy}(t) + \int \gamma(t-t')\dot{Q}_{xy}(t')\,\mathrm{d}t' + \omega_{v-l}^2 Q_{xy}(t) = 0, \qquad (2.46)$$

where we drop the upper tilde, $\tilde{}$, in order to simplify the mathematical notation. This equation enables the evaluation of time evolution of the Q variable, and hence of the OKE response. In fact, as we showed in (2.23), the OKE response function $R_{\mathrm{oke}}(t)$ is defined by the time derivative of the correlation function of the susceptibility that in the present case reduces to the orientational tensor $Q_{xy}(t)$, see (2.42). Alternatively, $R_{\mathrm{oke}}(t)$ can be directly calculated from the dynamic (2.46), through the Green function of this integro-differential expression, $G_{xyxy}^{QQ}(t)$. In fact, we have simply [9] $R_{\mathrm{oke}}(t) \propto G_{xyxy}^{QQ}(t)$.

The dynamic (2.46) is an integro-differential second-order expression where it appears a simple term proportional to Q and a complex term defined by the convolution of the Q time-derivative. The simple term describes a vibrational/librational dynamics summarized in the ω_{v-l} frequency. This dynamical process, similar to the previous "fast" dynamics, could be considered the oscillatory motion of a representative molecule in an effective local harmonic potential well defined by the frozen local liquid structure. The complex term describes the slower dynamical phenomena by means of a memory function, $\gamma(t)$. This function is the key parameter to describe the dynamics of complex liquids. This dynamical equation and the memory function can be also rigorously obtained from the Mori-Zwanzig theory [42,52]. Depending on the γ-function definition different dynamic phenomena can be included and described by the model. The simplest possibility is the Markovian approximation: $\gamma(t) = \nu\delta(t)$, which transforms (2.46) into the dynamical equation of a simple damped oscillator.

2.3.2.3   Mode-Coupling Models

Alternatively the hydrodynamic equations of motion can determine directly the time evolution of the correlation functions instead of the mesoscopic variables, according to the Mori-Zwanzig model [42,52]. These equations are the starting point of the Mode-Coupling Theories (MCT) [60–62].

In the *schematic model* of MCT [60–62], the dynamic equation of the density correlator, $\Phi_{\rho\rho}(t) = \langle\rho(t)\rho(0)\rangle/\langle\rho^2(0)\rangle$, is taken in the following form:

$$\ddot{\Phi}_{\rho\rho}(t) + \int \mathcal{M}(t - t')\dot{\Phi}_{\rho\rho}(t')\,\mathrm{d}t' + \Omega^2\Phi_{\rho\rho}(t) = 0, \qquad (2.47)$$

where the memory function $\mathcal{M}$ has a simple Markovian term, $\eta\delta(t)$, plus a real memory effect that, according to the schematic model, can be represented by a second-order expansion of the density correlator:

$$\mathcal{M}(t) = \eta\delta(t) + A\Phi_{\rho\rho}(t) + B\Phi_{\rho\rho}^2(t). \qquad (2.48)$$

So the MCT does not provide an analytical form of the memory function, as other theoretical models do, but it defines a general hierarchical way of building it. Clearly this definition of the memory introduces a self-coupling phenomena in the correlation dynamics. These coupled equations, (2.47) and (2.48), can be solved using a few asymptotic approximations. The solution of these equations provides an analytical description of the density dynamics, called *asymptotic results*.

The schematic MCT model predicts an unattained dynamical arrest of any density fluctuation at a critical temperature T_c. This *critical transition* has been related to the formation of a glassy state. In particular, the MCT model makes specific predictions for the dynamics of the density fluctuations above T_c and it provides a detailed analytical form for the density correlation function.[7] In the vicinity of T_c, the relaxation processes are asymptotically divided into two time scales: a fast/intermediate regime, called β-relaxation process, followed by a long time, or α-relaxation process. In the latter, the correlation function of the density fluctuations decays as a stretched exponential, often named the Kohlrausch-Williams-Watts law [60–62]:

$$\Phi_{\rho\rho}(t) = f\mathrm{e}^{-(t/\tau_\alpha)^\beta}, \qquad (2.49)$$

where τ_α is the structural relaxation time, β is the stretching factor, and f is the Debye-Waller coefficient. According to the MCT model, the structural times are strongly temperature dependent being $\tau_\alpha \propto (T-T_c)^{-\gamma}$, where γ is a critical exponent. In the intermediate time scale, i.e. in the β-regime, the relaxation can be described, as first approximation, by a double power law decay [60–62]:

$$\Phi_{\rho\rho}(t) = f + h_a t^{-a} - h_b t^b, \qquad (2.50)$$

the first power law is called critical decay and the second von Schweidler decay. The critical exponents a, b, and γ are linked by the following relation [60–62]:

$$\gamma = (1/2a + 1/2b). \qquad (2.51)$$

These predictions represent a complete dynamic scenario that has been investigated by numerous experiments [65]. Indeed, such MCT predictions have been able to describe properly the relaxation processes in supercooled liquids, even if the experimental results locate the MCT critical transition, T_c, at temperatures higher than the thermodynamical glass transition.

The OKE signal is dominated by the orientational dynamics (see (2.42)), hence the direct comparison of OKE response with the schematic model is based on the assumption of a complete similarity between density and rotational dynamics. This assumption seems valid in glass-formers for temperatures higher than the MCT critical transition. So according to this hypothesis, the OKE response function is [24]

$$R_{\text{oke}}(t) \propto -\frac{\partial}{\partial t}\Phi_{\rho\rho}(t), \qquad (2.52)$$

$$\alpha\text{-regime}: \quad R_{\text{oke}}(t) \propto (\beta/t)(t/\tau_\alpha)^\beta \, e^{-(t/\tau_\alpha)^\beta}, \qquad (2.53)$$

$$\beta\text{-regime}: \quad R_{\text{oke}}(t) \propto h_a \left[at^{(-a-1)}\right] + h_b \left[bt^{(b-1)}\right]. \qquad (2.54)$$

These MCT predictions have been compared with the OKE experiments on glass-formers [24–30, 37]. In these analyses, the authors found that the MCT asymptotic solutions properly reproduce the response relaxation in a large time–temperature range. Nevertheless for temperature $T \leq T_c$, the OKE relaxation shows several dynamic features not predicted by the schematic MCT model [27, 30].

Recently a more appropriate theoretical model has been used in order to describe the OKE data [36, 66]. This MCT model is called the *schematic two-correlator F_{12} model* [67] and it takes into account the rotational dynamics. The model is based on two dynamic equations. The introduced density equation (see (2.47) and (2.48)) represents the "master equation." A second equation, called the "slave equation", defines the rotational motions. This can be written defining the orientational correlator, $\Phi_{QQ}(t) = \langle Q_{xy}(t)Q_{xy}(0)\rangle/\langle Q_{xy}^2(0)\rangle$, and its memory equation:

$$\ddot{\Phi}_{QQ}(t) + \int m(t - t')\dot{\Phi}_{QQ}(t')\,dt' + \omega^2\Phi_{QQ}(t) = 0, \qquad (2.55)$$

being:

$$m(t) = \nu\delta(t) + C\Phi_{\rho\rho}(t)\Phi_{QQ}(t). \qquad (2.56)$$

The density dynamics does not depend by the orientational motion, as evident by (2.47) and (2.48), whereas the orientational dynamics clearly is determined by the density as is shown by (2.56). This translational-rotational coupling is present also when we are considering dynamic phenomena characterized by small or negligible q wave-vectors, different from the direct hydrodynamic coupling reported in [49] that is active only for large values of q.

The solutions of the F_{12} model are normally found by numeric calculations [36,66,67], because there does not exist a simple analytical solution. Only recently, the asymptotic solutions of this model have been calculated [66,68], providing a relatively simple functional form of the correlation functions.

According to (2.23) and (2.42), the OKE response function is defined by the time derivative of the orientational correlation function:

$$R_{\text{oke}}(t) \propto -\frac{\partial}{\partial t}\Phi_{QQ}(t) \qquad (2.57)$$

The comparison of F_{12} model with the OKE data has been performed only in a limited number of cases [36, 66] and it seems able to explain some of the complex relaxation features. We will discuss this shortly in Sects. 2.5.1.2 and 2.5.2.

2.4   The Experimental Procedure

In this section, we will describe the experimental apparatus that is typically utilized in order to perform a heterodyne-detected OKE experiment. In particular we will present the experimental systems developed and assembled in our laboratory.

First of all, a laser source able to produce short pulses with high energy is required. Presently there are a large variety of ultrafast laser systems available on the market. Substantially all these laser systems are based on the Ti:sapphire solid state active medium and on the Kerr lens mode-locking principle. There is a large amount of literature on this subject [69] and we will recall here only the principal features of our present laser system. This is composed of a Kerr lens mode-locked Ti:sapphire oscillator and a chirped pulse amplification system [19, 70], see Fig. 2.2. The oscillator source is a Kerr-lens mode-locked Ti:sapphire laser (FemtoRose 10MDC, by R&D Ultrafast Laser) pumped by

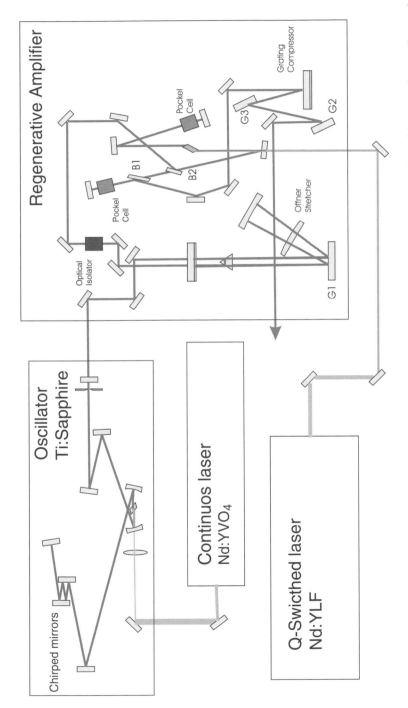

Fig. 2.2 Optical scheme of a typical laser system. Here we report an ultra-fast laser system based on Ti:sapphire solid state active medium and on the Kerr lens mode-locking principle. The main oscillator uses the chirped mirrors technique and the regenerative amplifier is based on a chirped pulse amplification scheme

the second harmonic of a laser-diode-pumped Nd:YVO$_4$ laser (Millennia V, by Spectra-Physics). Positive dispersion of a 2 mm thick gain medium is compensated by chirped mirrors. The round trip time is 13.7 ns corresponding to a repetition rate of 73 MHz. The minimum pulse width is around 10 fs for the large spectrum cavity operating mode. By a proper choice of the cavity chirped mirror the laser can operate in a narrower spectrum mode, giving a pulse width $\approx$20 fs. The amplification system (Pulsar, by Amplitude Technologies) is based on the chirped pulse amplification technique. The laser pulses from the oscillator are sent to an Offner stretcher and then to a Ti:sapphire regenerative cavity, pumped by a Q-switched Nd:YLF laser (Evolution, by Spectra-Physics). At the end, they are compressed back by a classical, two-grating, compressor design. The amplification stage produces $\approx$0.7 mJ at 1 KHz with a minimum pulse width of about 40 fs. The regenerative cavity includes two Pockel's cells in order to optimize the contrast ratio between main and satellite pulses (5×10^{-7}).

Though there is a most common optical scheme to perform HD-OKE experiments, a few variants are possible [71, 72].

Here we describe our optical set-up that has shown to be particularly successful in order to induce and detect the birefringence dynamics with high energy and low repetition laser pulses [19, 25, 27–29, 36, 37]. In this experimental set-up (see Fig. 2.3) the laser beam is divided into the exciting and the probing beam (about 80%–20%) by a beam splitter, BS1. The exciting pulse arrives

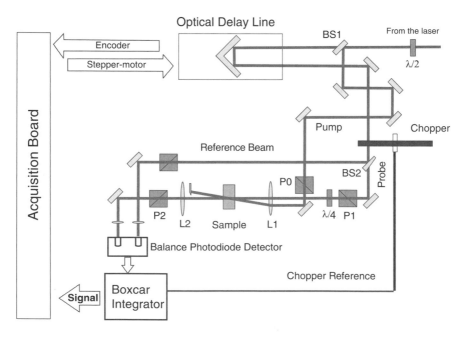

Fig. 2.3   A typical optical set-up and detection scheme used in OKE experiments

directly on the sample, passing through a polarizer P0, that produces a linear polarization along the $\hat{x}$ axis, see also Fig. 2.1. The probing pulse is delayed through an optical delay line, ODL, controlled by a computer. The ODL is made from a specular corner-cube, that produces a safe back reflection of the probe beam, mounted on a motorized stage. Typically, the time resolution of a OKE experiment is not determined by the limitation in optical delay control (easily better then 2–3 fs) but by the laser pulse duration. The probe then is split by a second beam-splitter (BS2) generating the so-called reference beam and finally it arrives at the sample through a polarizer P1 that selects the direction of polarization (45° with respect to the pump beam). Both the E^{ex} and E^{pr} are focused on the sample by a lens L1, typically of 500 mm focal length, which accomplishes the spatial superposition of the two pulses. A polarizer P2, crossed with respect to P1, is placed after the sample in order to select the polarization appropriate for the OKE experiment. To perform heterodyne detection (see next section) the signal field has to be beaten with an other field, called usually the local oscillator, E^{lc}. In this set-up this is obtained using a broad band $\lambda/4$ wave-plate, placed between P1 and the sample. This wave-plate is adjusted to get the minimum leakage out from the P2 polarizer, in the presence of the sample and of L1, then it is rotated by a small angle to generate the local field.

The detection of the signal in HD experiments is not always straightforward. In fact several contributions are simultaneously present on the detector, see next section, and the disentanglement of them is required. In our experiment we used a differential photodiodes system: one photodiode measures all the electromagnetic fields coming from the sample, $\mathbf{E}^{\mathrm{sg}} + \mathbf{E}^{\mathrm{lc}}$, the other photodiode is balanced with the reference beam to compensate for the intensity of the local field. The output of the differential is set to zero, in absence of the excitation beam. A boxcar gated integrator, synchronized to a mechanical chopper placed on the exciting beam, collects the output of the differential photodiode amplifier allowing an automatic subtraction of the other backgrounds present in the signal. The signal is at the end recorded by an acquisition board, which also controls and measures the position of the optical delay line. The homodyne contribution is eliminated by subtracting two different measurements taken with opposite direction of the rotation of the $\lambda/4$ plate. In fact, these two measurements are characterized by a positive terms for S_{homo} and by the S_{hete} signal with opposite signs, see (2.63).

The HD-OKE data are composed by a number of time-points corresponding to different positions of the optical delay line. When a step-by-step acquisition procedure [19] is used, the optical delay line moves one step and stops waiting for the signal acquisition. Typically the data have about few thousand of time-points. For each time-point the signal intensity is the average over few hundreds of laser pulses. If a very wide experimental time window is needed we perform separated scans of the delay line, characterized by a series of time-points spaced with different steps (e.g. from 0 ps to 4 ps we use a time-step of 5 fs, from 2 ps to 18 ps we use a time-step of 10 fs and for the remaining time window a logarithmic time-step). Then these scans are matched to rebuilt a single datafile, using an extreme attention to the signal consistency in the overlapping time regions.

We found extremely useful to perform each scan with optimized laser pulses [28, 37]. Typically pulses of different energy and temporal length are utilized to optimize the signal-noise ratio e.g., excitation pulses of 50 fs and 1.2 μJ (transform-limited) are employed in the short scan and excitation pulses of 1 ps and 12 μJ (chirped) in the long scan. With longer laser pulses, it is possible to transfer a larger quantity of energy to the system without increasing the instantaneous intensity of the pulse. The signal is then characterized by a better signal–noise ratio that allows access to a longer temporal scale. Using the heterodyne detection and stretched pulses more than five intensity decades can be scanned (see Fig. 2.4).

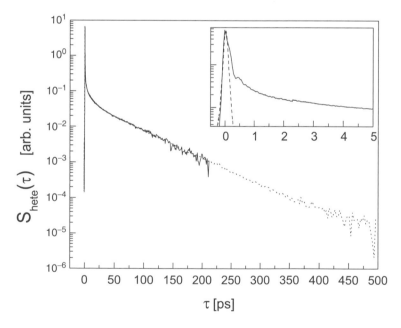

Fig. 2.4 The signal measured in a HD-OKE experiment on a epoxy resin (phenyl glycidyl ether) [28]. We show the OKE data obtained with laser pulses of different durations. In the long scan (dotted line) we used laser pulses of 1 ps, in the short scan (continuous line) the pulses were of 50 fs. In the inset figure, we show the same data in a very short time window. The dashed line is the instrumental response, obtained performing the OKE experiment on a sample with only instantaneous response, typically a quartz plate

2.4.1 Heterodyne Detection

In the heterodyne detection scheme, the signal, $\mathbf{E}^{\mathrm{sg}}$ defined in (2.21), interferes with an extra field, called local field $\mathbf{E}^{\mathrm{lc}}$, on the detector. The integrated intensity, measured by the main photodiode D1 in our set-up, is then

$$S_{\mathrm{tot}} \propto \int \mathrm{d}t \; |\mathbf{E}^{\mathrm{sg}}(t) + \mathbf{E}^{\mathrm{lc}}(t)|^2$$

$$= \int \mathrm{d}t \; \left\{ |\mathbf{E}^{\mathrm{lc}}(t)|^2 + |\mathbf{E}^{\mathrm{sg}}(t)|^2 + 2\mathrm{Re}[\mathbf{E}^{\mathrm{sg}}(t) \cdot \mathbf{E}^{\mathrm{lc}}(t)^*] \right\}. \quad (2.58)$$

There are three contributions in the previous equation: the first is the intensity of the local field, the second is the homodyne signal, see also (2.9), and the third one is the heterodyne signal

$$S_{\mathrm{lc}} \propto \int \mathrm{d}t \; |\mathbf{E}^{\mathrm{lc}}(t)|^2,$$

$$S_{\mathrm{homo}} \propto \int \mathrm{d}t \; |\mathbf{E}^{\mathrm{sg}}(t)|^2,$$

$$S_{\mathrm{hete}} \propto \int \mathrm{d}t \; 2\mathrm{Re} \left\{ \mathbf{E}^{\mathrm{sg}}(t) \cdot [\mathbf{E}^{\mathrm{lc}}(t)]^* \right\}. \quad (2.59)$$

Of course, in a HD-OKE experiment the relevant signal is the last one, the others are contributions to be eliminated. As we already mentioned, the S_{lc} signal can be read out with the differential photodiode detector. This is a fundamental experimental device and it allows us to clean up a strong contribution in the total signal, typically $S_{\mathrm{lc}}/S_{\mathrm{hete}} \sim 10-100$, which otherwise would seriously restrict the dynamic range of the detector. Furthermore, using as reference a part of the probe beam, the eventual effects of the slow drifts or fluctuations in the laser intensity are reduced. As we showed in the previous Sect. 2.2.1, the S_{homo} is indeed a signal that contains all the dynamical information (see (2.9)) but it is connected to the squared response. When the response has a complex relaxation pattern, this squared terms makes it even more difficult to address the different dynamic modes. So the S_{homo} contribution is considered a kind of spurious effect in the signal detection that has to be subtracted. This is possible because it does not dependent on the $\mathbf{E}^{\mathrm{lc}}$ and it can be subtracted performing two measurements characterized by opposite phase of the local field (i.e. π out-of-phase).

In our experimental set-up, the local field is generated by rotating the $\lambda/4$ plate of a small angle, ε. This wave-plate rotation produces a leakage of the probe beam from the P2 polarizer that represents our local field. This is proportional to the probe field and it can be calculated using the Jones matrix algebra [73]:

$$\mathbf{E}^{\mathrm{lc}} = (\hat{e}_x - \hat{e}_y)\eta \; E^{\mathrm{pr}} \text{ with } \eta = \varepsilon + i\varepsilon, \quad (2.60)$$

where ε is the angle formed between the fast axis of the wave-plate and the polarization direction selected by the polarizer of analysis, P2. Typically this is a small angle of few degrees and it can be positive or negative. The local field, so obtained, is a valid local field because it has an easily variable intensity and it is locked in phase with the signal field. If the local field is obtained by the wave-plate rotation the local field has a mixed phase (i.e. it contains in-phase and out-of-phase components) compared to the signal field, as is shown by (2.60). We can produce a local field also by rotating the P1 or the P2 polarizers. In such a case the local field has a well defined phase: the P1 rotation produces a $\pi/2$ out-of-phase field (i.e., $\eta = i\varepsilon$) and the P2 rotation produces a in-phase field (i.e., $\eta = \varepsilon$). As we described previously (see Sect. 2.2.2) if resonant effects are present, the induced polarization, and hence the signal field, has a different phase from the probe field. Considering that $E^{\text{sg}} \propto iP^3$, the part of the E^{sg} field that is in-phase with E^{pr} defines the dichroic effect, whereas the part that is $\pi/2$ out-of-phase describes the birefringence. So according to the definition of S_{hete} (see (2.59)) by rotating the P1 polarizer, we select the birefringence contribution and by rotating the P2 polarizer we select the dichroic. So, in principle, the creation of a local filed using the P1 and P2 polarizers has the advantage of a clear separation between birefringence and dichroic contributions. Indeed, we adopted this procedure for a direct experimental check of the presence of dichroic contribution in the OKE signal. Unfortunately, from the experimental point of view, any rotation of the polarizers produces a relevant optical mis-alignment, which makes this method complicated and not always reliable. For this reason is preferable to use the $\lambda/4$ wave-plate, which does not produce any modification of the optical alignment during its rotation.

As we have shown in (2.60), in our experimental procedure the local field has a mixed phase. So, if the response function has any resonances with the laser fields, the signal will have both birefringent and dichroic contributions, whereas in a nonresonant case (as we will consider in the derivation of (2.62)), the signal is determined only by the birefringence effect, even if a mixed phase is present in the local field. Indeed in our nonresonant experiments, we always found the dichroic signal absent or negligible.

According to the definition of the signal field, (2.21) and the local field expression (2.60), the heterodyne signal becomes

$$S_{\text{hete}}(\tau) \propto 2\text{Re}\left\{ i(\varepsilon - i\varepsilon) \int dt\, E^{\text{pr}}(t)E^{\text{pr}}(t)^* \int dt'\, R_{\text{oke}}(t - t')[E^{\text{ex}}(t')]^2 \right\}.$$

(2.61)

Since we are considering an OKE experiment the response function is a real number and this equation can be further simplified:

$$S_{\text{hete}}(\tau) \propto \varepsilon \int dt\, [\mathcal{E}(t - \tau)]^2 \int dt'\, R_{\text{oke}}(t - t')[\mathcal{E}^{\text{ex}}(t')]^2,$$

(2.62)

where we also introduced the definition of laser pulse envelope, (2.A.6), and the relative intensity expressions, according to (2.A.2). The derived equation

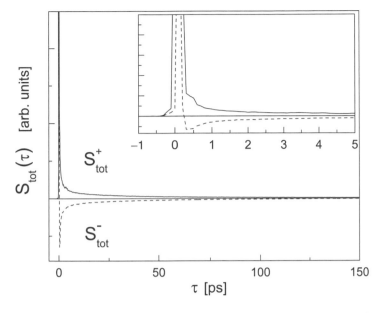

Fig. 2.5 HD-OKE experiment on glass-former m-toluidine at 273 K [25]. We show here the two measured signals $S_{\text{tot}}^{+}(\tau)$ and $S_{\text{tot}}^{-}(\tau)$ corresponding at two $\pm\varepsilon$ opposite values of the rotation angle of the wave-plate

shows how the heterodyne signal has a sign that depends on the angle ε defined by the wave-plate rotation, so that S_{hete} is positive or negative if ε is positive or negative. However the S_{homo} is always positive. By performing two measurements with opposite values of the angle $\pm\varepsilon$, we will get two opposite signal, S_{tot}^{+} and S_{tot}^{-}; it is immediately clear that we can extract from these total signals the heterodyne component:

$$\frac{1}{2}[S_{\text{tot}}^{+}(\tau) - S_{\text{tot}}^{-}(\tau)] \propto S_{\text{hete}}(\tau). \qquad (2.63)$$

Performing this subtraction we disentangle the heterodyne signal from the homodyne contribution and, furthermore, we clean up the signal from all the other spurious effects that are not sensitive to the phase of the local field. In Fig. 2.5 we report an experimental example of this procedure.

2.4.1.1 Recent Up-Grading

Recently we introduced an innovative acquisition system for HD-OKE spectroscopy [70]. It is based on a real-time acquisition of the experimental signal during the rapid scan of the optical delay line. This acquisition scheme enables data acquisition with high absolute time resolution (1 fs), high scanning velocity (2.5 cm s^{-1}) and long time delay (several ns). Moreover it is suitable

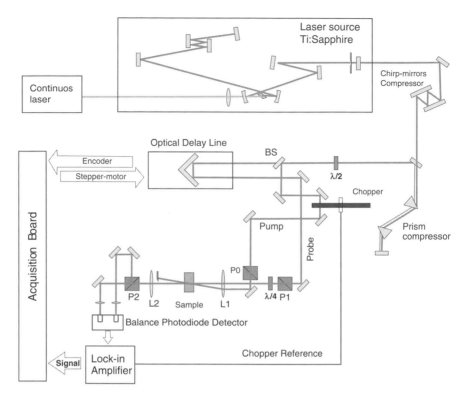

Fig. 2.6   New experimental set-up for HD-OKE spectroscopy with high repetition laser source and real-time data acquisition

for HD-OKE experiments performed with high and low repetition rate laser sources.

In Fig. 2.6 we report the experimental sketch of a HD-OKE experiment, recently built in our laboratory, employing the oscillator as laser source and the aforementioned acquisition system. The optical set-up is identical to that previously reported, apart from an additional optical compressor stage. The pulse compression is realized by the multireflection on two chirped mirrors and the passage through a pair of silica prisms. The background subtraction is obtained using a fast balanced photodiode detector and filtering the signal with a lock-in amplifier. In the real-time acquisition procedure, the delay line moves continuously and the signal is simultaneously acquired, while the delay line position is measured reading the encoder output. Such a system substantially decreases the acquisition time of each single delay scan, improving substantially the signal statistics.

Furthermore, we used the optical set-up in a different way [71]. In this scheme, the $\lambda/4$ wave-plate is rotated in order to produce a circularly polarized

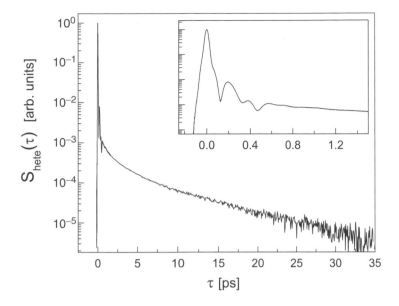

Fig. 2.7 HD-OKE signal on water at 249 K from the new experimental set-up

probe field and the two signals, corresponding to the horizontal and vertical linear polarizations generated by P2, are sent to the balanced photodiode detector. In such a way, an automatic heterodyne of the OKE signal is produced. A further reduction of the noise in the data can be obtained by performing a super-heterodyning of the OKE signal [72]. This is achieved by performing two OKE measurements, corresponding at left and right circular polarizations of the probe field, and making the difference of these two data.

In Fig. 2.7 we report HD-OKE signal on water obtained with this new experimental set-up and technique. Water has a very low OKE signal and it is a particularly difficult sample. Usually it requires quite a complex procedure to be properly investigated with the step-by-step acquisition [37]. With the real-time acquisition it is possible to get a very good signal/noise ratio with an extended dynamic range performing a single and continuous scan of the optical delay line without pulse energy/duration adjustments.

2.4.2 HD-OKE vs. Light Scattering Signal

The HD-OKE signal is directly connected with the response function (see (2.62)); in fact they are coincident when the laser pulse duration is much shorter than the dynamic time scale present in the response function (see (2.12)). Furthermore, the response function is defined by the time-derivative of the correlation function of the dielectric tensor (see (2.23)). These relationships permit the connection of the HD-OKE signal with that measured in a forward (i.e. the value of $q \sim 0$) depolarized light scattering experiment (DLS). In fact, a DLS experiment measures the spectral density of the dielectric tensor fluctuations [42], that according to the fluctuation-dissipation theorem can be defined as [43, 74]

$$S_{\mathrm{dls}}(\Delta\omega) \propto [n(\Delta\omega) + 1] \, Im\chi_{\mathrm{dls}}(\Delta\omega), \qquad (2.64)$$

where $\Delta\omega$ is the frequency difference between the incident and the scattered light, $n(\Delta\omega)$ is the occupation number, and $[n(\Delta\omega) + 1] = (1 - e^{-\frac{\hbar\Delta\omega}{K_B T}})^{-1} \simeq \frac{K_B T}{\hbar\Delta\omega}$ is the so-called Bose factor. $\chi_{\mathrm{ls}}(\Delta\omega)$ is the DLS frequency-dependent susceptibility corresponding to the Fourier transform of the time-dependent response function:

$$\chi_{\mathrm{dls}}(\Delta\omega) = \int e^{i\Delta\omega t} \, R_{\mathrm{oke}}(t) \, dt. \qquad (2.65)$$

Equations (2.64) and (2.65) show directly the connection between the OKE response function and the DLS susceptibility. So finally, we can easily connect the DLS and OKE signals:

$$S_{\mathrm{dls}}(\Delta\omega) \propto [n(\Delta\omega) + 1] \, Im \int e^{i\Delta\omega t} \, S_{\mathrm{oke}}^{\mathrm{hete}}(t) \, dt. \qquad (2.66)$$

This relation has been also checked experimentally [74, 75].

2.5   Some Experimental Results

There is a large amount of literature on OKE results, as we summarized in the introduction. A complete review of this literature is beyond the aim of this section. We simply summarize some of our OKE results on different complex liquids, trying to give a general scenario of the liquid dynamics as measured by OKE experiments. Furthermore, we do not discuss the vibrational part of the dynamics, which has been extensively reviewed in the first chapter. The OKE data reported in this section has been already published or will be shortly, and so we are here just recalling the more relevant aspects. For the details we refer to the published papers.

We divided the data of relative simple liquids, as benzene [36] and iodo-benzene [19], from the more complex liquid data, as super-cooled and glass-formers [25, 27]. A special attention is finally devoted to the dynamics of water [37].

2.5.1   HD-OKE in "Simple" Molecular Liquids

The liquid phase formed of small molecules (e.g. CS_2 and C_6H_6) has been largely studied by many OKE experiments [7,8,10,13,18–20,36] and numerical investigations [46,47,58,59]. Despite the relatively simple molecular structure, the dynamics of these liquids still present unclear aspects [20, 36, 59].

2.5.1.1 The Dynamics of Liquid Iodobenzene vs. Debye-Stokes-Einstein Model

As we already introduced in Sect. 2.3.2.1, the slower dynamics of simple liquids has been often described on the basis of the Debye-Stokes-Einstein (DSE) diffusion model. In our paper, Bartolini et al. [19], we checked such model on the dynamics of liquid iodobenzene as measured by HD-OKE experiments.

Actually, the interpretation based on such a single molecule picture appears as an oversimplified one. First of all, the observable in OKE experiments (as well as in a light scattering measurement) is collective in nature. On the other hand, the important role of collective motion and of cooperative effects has been demonstrated for the long time dynamics in several liquids characterized by strong anisotropic interactions. The importance of collective dynamics and of intermolecular interactions is also obvious in the short time limit: The picture of a single molecule librating in the potential well made up by the surrounding molecules, if acceptable for a "slow" solute in a "fast" solvent, is clearly oversimplified for neat liquids. In particular, it completely misses the key point of the correlated dynamics of neighboring molecules, which instead is of fundamental importance to understand the mechanism of liquid dynamics.

In Bartolini et al. [19] we presented the results of HD-OKE experiments performed on liquid iodobenzene in a wide temperature range. The experimental procedure employed in this work has been described in Sect. 2.4; other details can be found in the paper.

A typical long scan of the HD-OKE signal is shown in Fig. 2.8. At least three different time regimes can be distinguished in the relaxation pattern: (1) a long time scale, characterized by an exponential decay, with a time constant ranging between 10 and 20 ps, depending on the temperature; (2) an intermediate range, where the relaxation is quasi-exponential; and (3) the short time dynamics, appearing and lasting for a few picoseconds, showing a well-pronounced fast oscillations together with the early part of the relaxation process. Here we summarize the analysis of the slow and intermediate parts of the OKE signal, while the analysis of the faster dynamics can be found in the paper. According to the model introduced in Sect. 2.3.2.1, the slow and intermediate dynamics have been ascribed to rotational diffusion processes. A first attempt in order to extract the relaxation times present in the signal is the subtraction procedure. For the iodobenzene signal this is shown in Fig. 2.9.

The previous decomposition of the data relaxation suggests a double exponential form of the OKE response function and hence of the related correlator, $\Phi^{\text{oke}}(\tau) = A_1 e^{-\frac{\tau}{\tau_1}} + A_2 e^{-\frac{\tau}{\tau_2}}$. The two relaxation times, τ_1 and τ_2, are compatible with the DSE picture. As reported before, the diffusive relaxation is in general described by a multiexponential decay [42], which for an asymmetric rotator reduces to a biexponential if the principal axes of the rotational diffusion

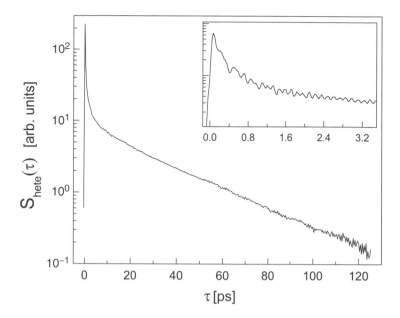

Fig. 2.8 HD-OKE signal from iodo-benzene sample at 273 K. In the inset we show the fast relaxation time scale, where several signal oscillations take place

tensor and of the molecular polarizability coincide. In the DSE model, as shown in (2.45), these relaxation times can be calculated using the shear viscosity, η, the molecular volume, V_{m}, and the parameter f. This last parameter depends on the hydrodynamic boundary conditions, slip or stick, and on the shape of the molecule.

The viscosity/temperature dependence of the τ_1 time constant, obtained from the fit of OKE data on iodobenzene, is in substantial agreement with the DSE model for a symmetric rotator within the slip condition. The slower time constant τ_1 coincides with the tumbling motion around the short molecular axis. This result is common to several molecular liquid dynamics. On the other hand, τ_2 is poorly reproduced by the DSE model. In fact if a prolate ellipsoid shape is assumed for iodobenzene, a symmetric rotator, τ_2 contains contributions from both spinning and tumbling reorientations characterized by time values much longer than those measured. If a more complete DSE model based on a asymmetric rotator is used to address the two relaxation times, the situation does not improve. Again τ_1 is correctly calculated but values of τ_2 are completely missed. In the faster time scale of the τ_2 relaxation the molecular motion is affected by the details of the short living local structure to a much larger extent, in comparison to the slower τ_1 dynamics, which instead is determined by interactions averaged over a much longer time.

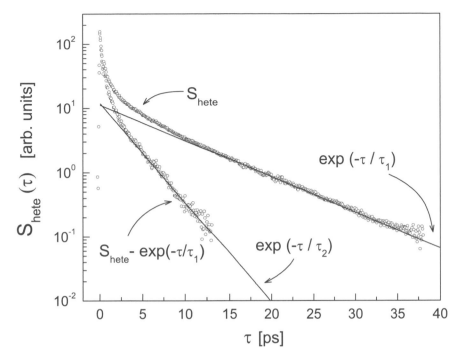

Fig. 2.9 The long part of HD-OKE signal, at 371 K, is well fitted by a single exponential decay of about 7 ps. If we subtract the long decay from the signal a second quasi-exponential decay appears characterized by a ≈ 2 ps relaxation time

This behavior is very similar to what was observed in other molecular liquids. These results on liquid iodobenzene, together with the other similar studies, pointed out clearly two aspects. Indeed the simple DSE model is able to describe quantitatively the slow relaxation processes. On the other hand, it is not able to properly address other simple dynamical features as the intermediate quasi-exponential relaxation. In our opinion, the DSE is really too simple an approach and the agreement observed for the slower relaxation could be partially fortuitous. Furthermore the starting hypothesis of decoupling of fast and slow dynamics does not have any safe physical background.

2.5.1.2 The Dynamics of Liquids Benzene vs. Mode-Coupling Theory

The dynamics of many molecular liquids, characterized by a simple molecular structure, show similar features. As we reported in the case of liquid iodobenzene, the experimental decays demonstrate the presence of three main time scales: (1) A *fast time scale* (typically up to a few picoseconds) where the driving processes are the intramolecular vibrations; the intermolecular dynamics on this time scale also have a vibrational character. In fact, on the short time scale the molecules are vibrating around their equilibrium position since it should be a valid approximation to take the liquid structure as static. (2) An *intermediate time scale* (typically some picoseconds) where the liquid structure cannot be taken as static since the diffusion processes start to be effective, producing a structural redefinition. On this time scale the intermolecular vibrations couple strongly with the structural relaxation rearrangement. (3) Finally, the *slow time scale* (whose range strongly depends on the liquid viscosity, varying from 10^{-11} to 10^{-9} s) where the leading effect is the structural relaxation. These recurrent features suggest a possible common dynamic scenario for molecular liquids. The DSE model has been largely utilized to address the slow relaxation. Indeed the apparent quantitative agreement with the data made this approach very popular even if its starting hypotheses are not physically accurate. The fast dynamics, intrinsically vibrational, has been described by many different models (e.g. the Kubo model of stochastic oscillators) with some success. However the nature of the intermediate relaxation remains obscure.

In Ricci et al. [36], we presented a study of the relaxation processes of the liquid and super-cooled benzene by HD-OKE spectroscopy. We have chosen benzene, consisting of perfectly symmetric molecules, as a particularly simple and meaningful example. The time-resolved optical Kerr effect data on benzene are obtained in a large temperature range, including the supercooled state (we succeeded in supercooling benzene by about 20 K below the melting point T_{m} = 279 K). Benzene is a highly symmetric molecule with a stiff molecular structure that does not present intramolecular dynamic contributions to the accessible OKE time scales. Thus it has proved to be an excellent model of a simple molecular liquid. The experimental results have been compared with the two-correlator F_{12} MCT model, see Sect. 2.3.2.3.

Our data [36] confirms and extends to the supercooled phase the results reported in the literature [15, 20]. The dynamics of benzene (see Fig. 2.10) shows three different processes: at very short delays (0.1–1 ps), there is evidence of an oscillatory behavior of the signal, superimposed as a fast quasi-exponential decay. No intramolecular vibration of benzene is low enough to be directly excited within the pulse bandwidth [27]; the origin of the oscillation is then to be found in the intermolecular vibrations. This oscillating signal is followed by a residual intermediate decay taking place up to 4–5 ps. Finally, the slow relaxation process appears and it extends, for the lower-temperature data, up to 40 ps. In several literature reports, the benzene slow dynamics has been attributed to the single molecule orientational diffusion process and it has been analyzed using the DSE model, similarly to our analysis of liquid iodobenzene previously reported. The fast and intermediate dynamics are more complex phenomena that still do not find an agreed interpretation [58, 59]. It has been compared with different models usually based on the local intermolecular vibrations, whose definition is generally quite arbitrary. According to these studies,

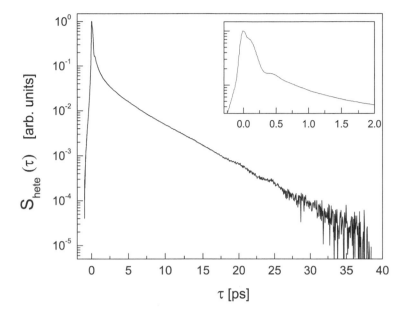

Fig. 2.10 HD-OKE signal from supercooled benzene sample at 269 K, about 10° below the melting point. In the inset we show the fast relaxation time scale, where a smooth signal oscillations take place

two main groups of vibrational modes are present: the first group is made up of under-damped modes characterized by an approximate frequency of 1.2 THz and the second group consists of over-damped modes of about 360 GHz. The first modes describe the fast vibrational dynamics; the second modes reproduce the intermediate relaxation processes. These models require a substantial decoupling between the slow/diffusive dynamics and the fast-intermediate processes. Such a hypothesis is in many respects an oversimplification of the dynamic problem that hardly applies to a molecular liquid such as benzene where all the dynamic features are characterized by similar time scales. Thus a definite assignment of the measured decay has not been worked out. In particular, the nature of the intermediate relaxation remains obscure. Since these dynamic features are repeated and common to many molecular liquids, the open questions are particularly relevant.

We investigated MCT as a possible valid model to describe the dynamics of the molecular liquids. Indeed, MCT models have been able to describe the relaxation features of several glass-formers up to temperatures well above the critical temperature [24, 25, 27–29, 37]. This suggests that the MCT should be tested as a possible model for the dynamics of molecular liquids, even if they do not show a glass transition and are characterized by rather low viscosities. We utilized the schematic two-correlator F_{12} MCT model in order to describe

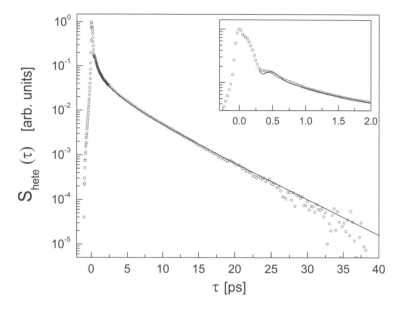

Fig. 2.11   The HD-OKE signal, at 269 K, is well fitted by MCT model up to ≈ 0.5 ps time

the OKE signal on benzene. The integro-differential equations of the model, reported in Sect. 2.3.2.3, have been solved numerically adjusting the A, B, C, Ω, and ω parameters thorough a fitting procedure in order to optimize the time derivative of the orientational correlator (see (2.57)) to the measured data. A typical fit is reported in Fig. 2.11.

The numerical solution of this model is indeed able to properly fit the benzene decay in the whole relaxation time scale, from about 0.5 to 35 ps. What is striking is the ability to fit the fast, intermediate and slow dynamics without any arbitrary separation of the time scales and of the dynamic processes.

In this MCT model, all the dynamic processes are intrinsically included and coupled. The fast vibration, the slow structural relaxation, and the intermediate decay appear as solutions of the equations of motion. The vibrational dynamics is characterized by two frequency parameters, Ω and ω, of about 1 THz, in reasonable agreement with the mean frequency of the vibrational modes found in previous studies. The crossover from the fast to the slow processes correctly describes the intermediate relaxation. This model does not require extra over-damped vibrational modes, as required by the previous interpretations. According to this interpretation, the intermediate relaxation is simply the merging of the local vibrational dynamics into the slow collective diffusive process. This merging is not trivial, as it has already been proved in glass-former liquids, because the two processes are strongly coupled.

In our opinion, the MCT interpretation represents a valid physical model for the molecular liquids, as here shown for the benzene liquid, and for glass-formers, as we are going to show in the next section.

2.5.2 HD-OKE in Supercooled Liquids and Glass-Formers

Recently the OKE experiments have been shown to be a particularly useful spectroscopic tool to investigate glass-former dynamics [24–30]. Indeed the OKE data enabled a better check of the theoretical models, especially of MCT approaches. In this Section, we summarize the studies of the m-toluidine glass-former performed by the authors [25,27] as a particularly meaningful example.

Meta-toluidine (CH_3–C_6H_4–NH_2) is a disubstituted benzene ring with CH_3 and NH_2 groups in the 1, 3 positions; it is one of the simplest fragile liquids, which remains very easily supercooled to its thermodynamic glass transition temperature $T_g = 187$ K, while the melting temperature is $T_m = 243.5$ K. Starting with 99% pure m-toluidine purchased from Merck, the sample was purified by distillation under vacuum and then kept in a quartz cell. The cell was placed in a cryostat system, cooled with Peltier cells; this enabled temperature control to better than 0.1 K. Our HD-OKE measurements were performed at 15 temperatures between 295 and 225 K with the experimental equipment and laser system, presented in Sect. 2.4. Let us simply recall here that the time resolution allows us to measure the response function by a step-by-step procedure, up to approximately 4 ns. The amplitude of the HD-OKE signal may be detected over a dynamic range of approximately five decades and exhibits a complex relaxation pattern.

In Fig. 2.12 we show HD-OKE signal data on a log–log scale. Each datum is an average of five independent measurements. The measured dynamics exhibit three different regimes: a nearly temperature-independent region from 0.1 to 2–3 ps, where oscillations related to intramolecular vibrations appear; an intermediate decay region from 2 to about 10 ps, depending on the temperature; and the final long-time relaxation. The long-time dynamics have a stretched-exponential form with decay times increasing with decreasing temperature. All the data sets have the same general form, but these major features (the three regimes) occur on different time scales depending on the temperature.

In the supercooled phases of glass-former liquids, the rotational dynamics is expected to be strongly coupled with the density dynamics. Therefore, the orientational and translational dynamics should present similar features. This sensible hypothesis has been verified by several numerical and experimental results [65]. In this framework, it is worthwhile to compare the OKE data on rotational dynamics to the predictions of the schematic MCT model on density dynamics. This comparison is particularly meaningful because the

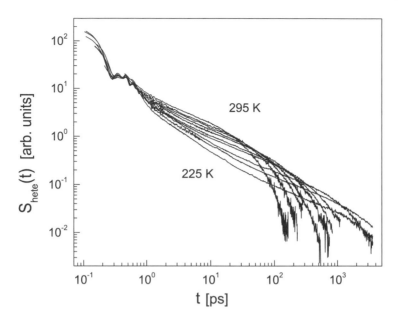

Fig. 2.12   The HD-OKE signal from liquid and supercooled m-toluidine sample. We report the data in a log–log plot in order to show clearly the different time scales present in the relaxation process

asymptotic results, introduced in Sect. 2.3.2.3, define a clear and not trivial dynamical scenario.

Indeed, the MCT asymptotic results have shown to describe successfully the relaxation of glass-formers for relatively weak supercooling condition [65] (i.e. for relatively high temperature, $T \geq T_c$). Whereas for deep supercooling condition (i.e. $T \leq T_c$) the MCT asymptotic results are not applicable, since they predict a structural arrest at the critical temperature (i.e. $\tau_\alpha \to \infty$ when $T \to T_c$), which is not experimentally verified in glass-formers. Therefore, more complex MCT models must be used in order to reproduce the experimental results. The OKE experimental results have proved to be a critical check for the asymptotic MCT predictions [24–30]. In particular, we analyzed the intermediate and slow relaxation of OKE signal on m-toluidine according to the asymptotic solution of the schematic MCT model [25]. The fast dynamics has been partially compared with more complex MCT approaches, as the two-correlator F_{12} model [27]. Here we report only the study of the slower relaxation processes.

As shown in Sect. 2.3.2.3, the schematic MCT model can be solved using a few asymptotic approximations [60–62], providing a relatively simple analytical expression for the measured response function (see (2.53) and (2.53)). According to these equations, the analysis of the slow relaxation vs. α-regime can be based on a master plotting technique. In fact, following (2.53) all α-correlation

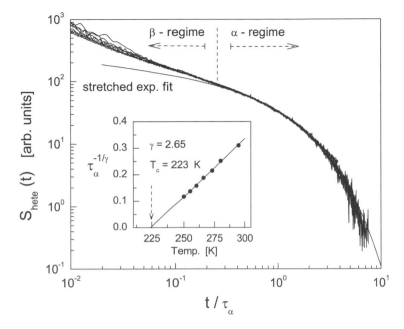

Fig. 2.13 The HD-OKE signals at different temperatures are reported in a master-plot. The time scale of each decay has been divided by a proper τ_α time in order to have the largest overlap possible. A fit according to the derivative of a stretched exponential is shown with $\beta = 0.8$. In the inset, we report the linearized plot of the structural relaxation times vs. temperature

functions should collapse in a single master-function when reported in a rescaled time axis, t/τ_α. In Fig. 2.13, we report the master-plot of m-toluidine data together with temperature dependence of the extracted structural relaxation times. The figure clearly shows that the MCT time–temperature superposition principle holds in the α-regime, enabling a good overlap of OKE data. The linear dependence of structural relaxation times, shown in the inset, proves the validity of the critical law, $\tau_\alpha \propto (T - T_c)^{-\gamma}$, and its fit gives the values of T_c and γ critical parameters.

The intermediate dynamics can be compared with asymptotic MCT predictions of the β-regime. In Fig. 2.14 we report only three OKE plots corresponding to the lower temperatures with the best fit of β-response function, see (2.54). The critical parameters a and b are not temperature dependent and linked by (2.51) to the γ exponent, obtained by the α-regime analysis. The OKE data are well reproduced by the MCT β-correlator for temperature higher than 235 K, corresponding to $T_c + 10$ K, but for lower temperatures a different relaxation appears in the 2–20 ps time window. This has been addressed, in other glass-formers, as a nearly-logarithmic decay since it would be connected to a logarithmic decay of the correlator [66, 68]. In fact, if $\Phi(t) \approx \log(t)$ we will have $R_{oke}(t) \approx t^{-1}$. In our m-toluidine data, we found some evidence of a possible extra power law decay, as shown in Fig. 2.14, with an exponent $c \approx 0.85$.

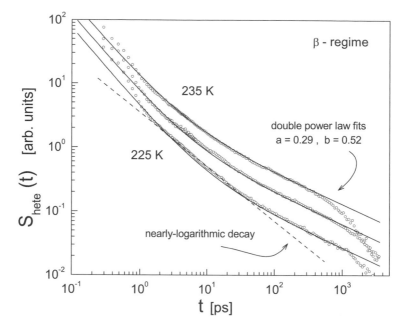

Fig. 2.14   The HD-OKE signals at lower temperatures (225, 229, and 235 K) are reported with a fit according to the derivative of the MCT double power law decay with $b = 0.52$ and $a = 0.29$. An arbitrary vertical shift has been introduced in order to avoid data overlap. In the data decay at 225 K a clear departure from the MCT prediction appears, the presence of a possible nearly-logarithmic decay is evidenced by the dashed line corresponding to a t^{-c} decay with $c \approx 0.85$

   In conclusion, we found that the predictions of the schematic MCT model are in very good agreement with the OKE data for weak supercooling but the agreement disappears approaching the critical temperature.

   The general experimental scenario of glass-formers for deep supercooling condition is still not clear and there is a live debate on the theoretical models. Concerning the OKE data, this issue has been partially investigated using the schematic two-correlator MCT model [66, 68].

2.5.3   The Water Case

We summarize here our results on supercooled water dynamics by HD-OKE experiments [37]. These results are relevant for two main reasons. First, because they deal with liquid water, one of the most relevant substances in the universe, that presents many anomalous chemical–physical properties. Second, liquid water represents a model system for the investigation of a liquid–liquid phase transition hypothesis.

   Many thermodynamic proprieties (such as heat capacity, compressibility, and thermal expansion coefficient) and dynamic properties (such as the shear viscosity, self-diffusion coefficient, and relaxation times) of water all exhibit an anomalous increase upon cooling [53]. They show a temperature dependence characterized by a diverging power law of the type $(T - T_s)^{-x}$, with $T_s \approx$ 223–228 K. The nature and physical origin of this singularity are still debated, but there are indications of structural modifications occurring around this temperature and the simulation results show that it could be related to the MCT critical temperature.

OKE experiments, as shown in the previous sections, can provide an accurate measure of the relaxation dynamics of molecular liquids, but the intensity of the OKE water signal is very low, making it very difficult to determine its relaxation profile accurately at times longer than a few picoseconds. Here we present the results of an extensive investigation by means of heterodyne detected OKE experiments (HD-OKE) of the relaxation in liquid and supercooled water, using the experimental set-up described in Sect. 2.4. We used optimized laser pulses in order to improve the experimental sensitivity. The entire time decay scan was divided into three sections: short laser pulses (60 fs) were used for the short time scan; for the intermediate time delay intervals the laser pulse duration was stretched up to 120 fs; for the longer time we used stretched laser pulses of 500 fs duration. In this procedure, we increased the pulse energy to keep the peak power roughly constant. The low noise level and the large dynamic range provide data of sufficiently high quality to extend measurements by almost 20° into the supercooled region, up to 254 K. A typical measured signal is shown in Fig. 2.15.

We provided the first unambiguous measurement of the entire correlation function in liquid water as a function of temperature. These data enabled a MCT analysis of the slow dynamic regime. Hence we compared our data with the MCT prediction in the α-regime as already shown for the glass-former liquids.

In Fig. 2.16, we collect all the data measured at different temperatures (314–254 K) in a log–log master-plot. The perfect coincidence of the different curves demonstrates that the slow relaxation of water follows a unique decay law in the entire temperature range considered. We found, for all temperatures, that the time-derivative of the stretched exponential function allows a very satisfactory fitting of the HD-OKE data. The stretching parameter β has a constant value of 0.6 throughout the entire temperature range considered.

As shown in the inset of Fig. 2.16, the temperature dependence of the measured relaxation time τ_α is well reproduced by the MCT power law, $(T - T_c)^{-\gamma}$, with $T_c = 221$ K and $\gamma = 2.2$. So the water singularity temperature $T_s \approx 225$ K, obtained from measurements of other dynamic and transport properties, substantially coincides with the MCT critical temperature.

Two important new results are obtained from our OKE experiment on supercooled water: (1) The stretched exponential nature of the relaxation and the

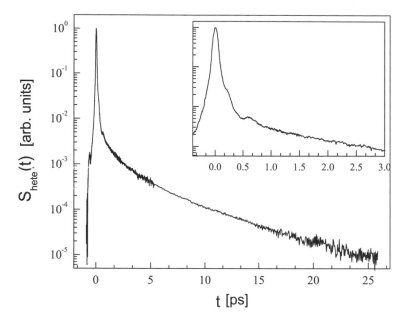

Fig. 2.15   Signal decay measured in the time-resolved HD-OKE experiment on supercooled water at $T = 257$ K. In the inset we reported the signal in the short time region. An oscillatory component occurs in the decay trace for time shorter than 1 ps; for longer times the relaxation shows a monotonic decay

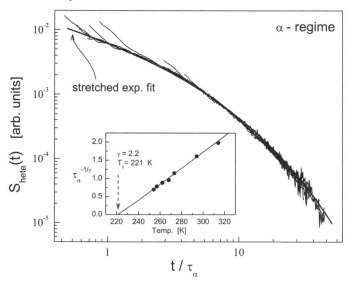

Fig. 2.16   Master plot obtained by rescaling the time and intensity axes of the HD-OKE signals. The inset shows the temperature dependence of the water relaxation times, obtained by the fit of HD-OKE signal decays with the time-derivative of the stretched exponential function

invariance of the stretching parameter with the temperature. The large stretching effect ($\beta = 0.6$) indicates that water dynamics is characterized by a distribution of different timescales, suggesting the presence of a variety of relaxing structures whose distribution does not, according to our results, change notably with temperature. (2) Our data also allow for a comprehensive comparison with the predictions of the MCT model. The MCT predictions agree with our experimental observations, both in terms of the form of the correlation function and in terms of its scaling with temperature.

A purely dynamic water model, as the MCT theory, clearly cannot account for the rich and anomalous thermodynamic behavior of water, which calls for other model types such as the much debated liquid–liquid transition hypothesis [53]. Nevertheless, these OKE results provide with unambiguous evidence that the anomalous behavior of weakly supercooled water at atmospheric pressure can successfully be described by a fully dynamic model, with no need for a thermodynamic origin for the observed anomalous behavior. In this picture the singularity temperature T_s, inferred from a range of different dynamic and transport properties of water, is the signature of an avoided dynamical arrest. Despite its strongly hydrogen-bonded nature, weakly supercooled water behaves in this context like a fragile glass-former, similar to many other molecular liquids.

Acknowledgments

The authors wish to thank R. Righini for his fundamental contribution in setting, performing, and understanding of the HD-OKE experiments. We are grateful to R. M. Pick for his continuous support in the glass-former studies. We would like to thank J. Palmer for his critical reading. We acknowledge the scientific encouragement by F. Sciortino and G. Ruocco. Special thanks goes to the technical staff of LENS. The presented research has been performed at LENS and it was also supported by EC grant N.RII3-CT-2003-506350, by CRS-INFM-Soft Matter (CNR), and MIUR-COFIN-2005 grant N. 2005023141-003.

Appendix: Laser Pulses Definition

The laser pulses typically used in an OKE experiment are characterized by an optical frequency (700 nm), a duration longer than 10–20 fs and they are normally transform-limited (i.e., with no time modulation in phase, apart from the optical frequency). The time dependence of the electric field generated by these laser pulses can be safely described as a plane wave in the approximation of the slow parameter variation, see [40]. Hence, using the complex notation we can write the expression of a linearly polarized pulse, as follows

$$\mathbf{E}(\mathbf{r}, t) = \hat{e}_i E_i(\mathbf{r}, t) = \hat{e}_i \mathcal{E}(\mathbf{r}, t) \, e^{i(\mathbf{k} \cdot \mathbf{r} - \omega t)}, \tag{2.A.1}$$

where $\hat{e}_i$ is a unit vector indicating the direction of polarization, $\mathbf{k}$ is the wave-vector defining the direction of propagation, ω is the optical frequency, and $\mathcal{E}(\mathbf{r}, t)$ defines the pulse envelope. $\mathcal{E}$ is the real, slowly varying time-dependent amplitude defining the pulse time duration, in particular for a continuous laser $\mathcal{E}$ is not time-dependent. Furthermore, it defines the intensity time dependence, since in the complex notation the pulse intensity is given by

$$I(t) \propto |\mathbf{E}(\mathbf{r}, t)|^2. \tag{2.A.2}$$

In the experimental configuration defined in Fig. 2.1, the linear polarization of the pump and probe are in the $\hat{x}$ directions and $45°$ in the $\hat{x}\hat{y}$ plane, respectively. Both the pulses are taken collinear propagating in the $\hat{z}$ direction.

$$\mathbf{E}^{\mathrm{ex}}(z, t) = \hat{e}_x \mathcal{E}^{\mathrm{ex}}(z, t)\, e^{i\omega(\frac{\eta}{c}z - t)}, \tag{2.A.3}$$

$$\mathbf{E}^{\mathrm{pr}}(z, t) = \hat{e}_x E_x^{\mathrm{pr}}(z, t) + \hat{e}_y E_y^{\mathrm{pr}}(z, t) = (\hat{e}_x + \hat{e}_y)\mathcal{E}^{\mathrm{pr}}(z, t)\, e^{i\omega(\frac{\eta}{c}z - t)}. \tag{2.A.4}$$

Here we consider the unique refractive index η, because we are considering two laser beams having the same frequency, ω, and thus propagating with the same phase velocity, $\frac{c}{\eta}$. Furthermore, in these equations we take simply an isotropic index of refraction, considering the beams propagating in the air before the sample, of course the beam polarization does not change during its propagation. For the excitation pulse this applies also during the sample propagation while it does not apply for the probe beam that finds the sample excited, hence an anisotropic index of refraction.

In a typical experiment the probe and excitation pulses are derived from a single laser beam split in two parts of different intensity. Hence the temporal and phase characteristics of the pulses are identical and defined by the main laser beam. Furthermore, in a time-resolved OKE experiment, the probe pulse is delayed by an optical delay that introduces a delay time $\tau = \Delta z \frac{\eta}{c}$, where Δz is the extra length introduced in the probe path. Hence we can write the envelope functions as

$$\mathcal{E}^{\mathrm{ex}}(z, t) = (1 - a)\ \mathcal{E}\left(\frac{\eta}{c}z - t\right), \tag{2.A.5}$$

$$\mathcal{E}^{\mathrm{pr}}(z, t) = a\mathcal{E}\left(\frac{\eta}{c}(z + \Delta z) - t\right) = a\mathcal{E}\left(\frac{\eta}{c}z - (t - \tau)\right), \tag{2.A.6}$$

where a is defined by the beam-splitter and the leading envelope function, $\mathcal{E}(t)$, by the main laser and by the group velocity compensation optics. Indeed, the weak anisotropy, induced by the excitation pulse, can be neglected in the envelope propagation (but not in the phase modification). Furthermore this means that the birefringence will be induced and probed with the same time sequence, making the experimental characterization easier [9].

Finally, the laser pulses reaching the sample can be described by the following equations:

$$\mathbf{E}^{\mathrm{ex}}(z, t) \simeq \hat{e}_x(1 - a)\mathcal{E}\left(\frac{\eta}{c}z - t\right)\, e^{i\omega(\frac{\eta}{c}z - t)}, \tag{2.A.7}$$

$$\mathbf{E}^{\mathrm{pr}}(z, t) \simeq (\hat{e}_x + \hat{e}_y)\, aE^{\mathrm{pr}}(z, t)$$

$$\simeq (\hat{e}_x + \hat{e}_y)\, a\mathcal{E}\left(\frac{\eta}{c}z - (t - \tau)\right)\, e^{i\omega(\frac{\eta}{c}z - (t - \tau))}. \tag{2.A.8}$$

Notes

1. The effect of polarizers and other active optical components, on the beam tracking can easily described by the Jones Matrix [73]. According to this picture the two component of the beam, E_x^{pr} and E_y^{pr}, are modified by the $-45°$ polarizer matrix, as following:

$$\left(\begin{array}{cc} E_x^{\text{sg}} & E_y^{\text{sg}} \end{array} \right) = \left(\begin{array}{cc} (E_x^{\text{pr}} - E_y^{\text{pr}}) & (-E_x^{\text{pr}} + E_y^{\text{pr}}) \end{array} \right) = \left(\begin{array}{cc} 1 & -1 \\ -1 & 1 \end{array} \right) \left(\begin{array}{c} E_x^{\text{pr}} \\ E_y^{\text{pr}} \end{array} \right)$$ where the

notation $(E_1 E_2) \equiv E_1 \hat{e}_x + E_2 \hat{e}_y$

2. In fact we have $\Delta \eta_{ij}(t) \propto \Delta \chi_{ij}(t) \propto \left(\begin{array}{cc} \int \mathcal{R}_{xxxx} |E^{\text{ex}}|^2 & 0 \\ 0 & \int \mathcal{R}_{yyxx} |E^{\text{ex}}|^2 \end{array} \right)$

3. The susceptibility tensor is the proper variable to describe the light-matter interactions, since it links directly the electromagnetic field and the polarization effects. Nevertheless, other variables such as the index of refraction or the dielectric function are very often used in the literature. The Maxwell equations provide the connection between the dielectric tensor, ϵ_{ij}, and the linear susceptibility tensor, χ_{ij} (in literature sometimes called liquid polarizability and reported as Π_{ij}): $\epsilon_{ij} = 1 + 4\pi \chi_{ij}$ in the CGS unit system. It is also straightforward the connection with the index of refraction, since $\eta = \sqrt{\epsilon}$. So when only the fluctuating or time-dependent part is relevant, the three functions are substantially equivalent: $\delta \chi_{ij} \propto \delta \epsilon \propto \delta \eta$

4. This polarizability corresponds to the isolated molecular polarizability normalized over the mean field interaction present in the liquid. It can be substantially different from the molecule polarizability measured in the gas phase.

5. If the pulse is transform limited [40] its spectrum and time profile are directly connected by a Fourier transformation. Hence the spectrum bandwidth is inversely proportional time duration: $\Delta \omega \propto \frac{1}{\Delta \tau}$. For a pulse of 100 fs duration, 100×10^{-15} s, we get a bandwidth of about 30 THz, equivalent to $100 \, \text{cm}^{-1}$. This spectrum width is often large enough to include some intramolecular vibration modes.

6. This expression corresponds to (2.9) at page 172 of the first paper by Pick et al. [49]. Here we neglect the last term of this equation that takes into account the coupling between rotational and slow translational dynamics. This approximation is valid for the limit $q \rightarrow 0$, as explicitly shown in Appendix C of that paper.

7. Here we report the simplified version of the MCT asymptotic solutions for the density correlator. They are obtained using the leading-order asymptotic solutions of the MCT. In the α-regime the scaling-law, or *time-temperature superposition principle*, is evident by (2.49). Whereas in the β-regime we do not make explicit the scaling-law that is hidden into (2.50). The rigorous derivation and the full expression of the MCT functions can be found in [60–62]. An experimental comparison between the full β-correlator expression and the next-to-leading order corrections with the OKE signal can be found in [25].

References

[1] Kerr, J. (1875). A new relation between electricity and light: Dielectrified media birefringent. *Philos. Mag.* 50: 336–348, ibidem 446–458. See also Kerr bibliography on (1907) *Nature* 76: 575–576.

[2] Mayer, G. and Gires, R. (1963). The effect of an intense light beam on the index of refraction of liquids. *C.R. Acad. Sci.* 258: 2039.

Marker, P.D., Terhune, R.W. and Savage, C.W. (1964). Intensity-dependent changes in the refractive index of liquids. *Phys. Rev. Lett.* 12: 507–509.

[3] Dugay, M.A. and Hansen, J.W. (1969). An ultrafast light gate. *Appl. Phys. Lett.* 15: 192–194.

Shimuzu, F. and Stoicheff, B.P. (1969). Study of the duration and birefringence of self-trapped filaments in CS_2. *IEEE J. Quantum Electron.* QE-5: 544–546.

[4] Levenson, M.D. and Eesley, G.L. (1979). Polarization selective optical heterodyne detection for dramatically improved sensitivity in laser spectroscopy. *Appl. Phys.* 19: 1–17.

[5] Waldeck, D., Cross, A.J., McDonald, D.B. and Fleming, G.R. (1981). Picosecond pulse induced transient molecular birefringence and dichroism. *J. Chem. Phys.* 74: 3381–3387.

[6] Greene, B.I. and Farrow, R.C. (1983). The subpicosecond Kerr effect in CS_2. *Chem. Phys. Lett.* 98: 273–276.

Greene, B.I., Fleury, P.A., Carter Jr., H.L. and Farrow, R.C. (1984). Microscopic dynamics in simple liquids by sub-picosecond birefringence. *Phys. Rev. A* 29: 271–274.

[7] Kalpouzos, C., Lotshaw, W.T., Mc Morrow, D. and Kenney-Wallace, G.A. (1987). Femtosecond laser-induced Kerr responses in liquid CS_2. *J. Phys. Chem.* 91: 2028–2030.

McMorrow, D., Lotshaw, W.T. and Kenney-Wallace, G.A. (1988). Femtosecond optical Kerr studies on the origin of the nonlinear responses in simple liquids. *IEEE J. Quantum Electron.* QE-24: 443–454.

[8] Lotshaw, W.T., McMorrow, D., Kalpouzos, C. and Kenney-Wallace, G.A. (1987). Femtosecond dynamics of the optical kerr effect in liquid nitrobenzene and chlorobenzene. *Chem. Phys. Lett.* 136: 323–328.

Kalpouzos, C., Lotshaw, W.T., McMorrow, D. and Kenney-Wallace, G.A. (1987). Femtosecond laser-induced Kerr responses in liquid carbon disulfide. *J. Phys. Chem.* 91: 2028–2030.

Kalpouzos, C., McMorrow, D., Lotshaw, W.T. and Kenney-Wallace, G.A. (1989). Femtosecond laser-induced optical Kerr dynamics in CS_2/alkane binary solutions. *Chem. Phys. Lett.* 155: 240–242.

McMorrow, D. and Lotshaw, W.T. (1990). The frequency response of condensed-phase media to femtosecond optical pulses: Spectral-filter effects. *Chem. Phys. Lett.* 174: 85–94.

McMorrow, D. and Lotshaw, W.T. (1991). Dephasing and relaxation in coherently excited ensembles of intermolecular oscillators. *Chem. Phys. Lett.* 178: 69–74.

McMorrow, D. and Lotshaw, W.T. (1993). Evidence for low-frequency ($15\,\mathrm{cm}^{-1}$) collective modes in benzene and pyridine liquids. *Chem. Phys. Lett.* 201: 369–376.

McMorrow, D., Thantu, N.J., Melinger, S., Kim, S.K. and Lotshaw, W.T. (1996). Probing the Microscopic Molecular Environment in Liquids: Intermolecular Dynamics of CS_2 in Alkane Solvents. *J. Phys. Chem.* 100: 10389–10399.

[9] Yan, Y. and Nelson, K.A. (1987). Impulsive stimulated light scattering. I. General theory. *J. Chem. Phys.* 87: 6240–6256.

Yan, Y. and Nelson, K.A. (1987). Impulsive stimulated light scattering. II. Comparison to frequency-domain light-scattering spectroscopy. *J. Chem. Phys.* 87: 6257–6265.

[10] Ruhman, S., Williams, L.R., Joly, A.G., Kohler, B. and Nelson, K.A. (1987). Nonrelaxational inertial motion in carbon disulfide liquid observed by femtosecond time-resolved impulsive stimulated scattering. *J. Phys. Chem.* 91: 2237–2240.

Ruhman, S., Joly, A.G. and Nelson, K.A. (1988). Coherent molecular vibrational motion observed in the time-domain through impulsive stimulated Raman scattering. *IEEE J. Quantum Electron.* QE-24: 460–469.

Kohler, B. and Nelson, K.A. (1992). Femtosecond impulsive stimulated light scattering from liquid carbon disulfide at high pressure: Experiment and computer simulation. *J. Phys. Chem.* 96: 6532–6538.

[11] Deeg, F.W., Stankus, J.J., Greenfield, S.R., Newell, V.J. and Fayer, M.D. (1989). Anisotropic reorientational relaxation of biphenyl: Transient grating optical Kerr effect measurements. *J. Chem. Phys.* 90: 6893–6902.

Greenfield, S.R., Sengupta, A., Stankus, J.J., Terazima, M. and Fayer, M.D. (1994). Effects of local liquid structure on orientational relaxation: 2-Ethylnaphthalene, neat and in solution. *J. Phys. Chem.* 98: 313–320.

[12] Palese, S., Schilling, L., Miller, R.J.D., Staver, P.R. and Lotshaw, W.T. (1994). Femtosecond optical Kerr-effect studies of water. *J. Phys. Chem.* 98: 6308–6316.

Palese, S., Mukamel, S., Miller, R.J.D. and Lotshaw, W.T. (1996). Interrogation of vibrational structure and line broadening of liquid water by Raman-induced Kerr-effect measurements within the multimode Brownian oscillator model. *J. Phys. Chem.* 100: 10380–10388.

[13] Waldman, A., Banin, U., Rabani, E. and Ruhman, S. (1992). Temperature dependence of light scattering from neat benzene with femtosecond pulses: Are we seeing molecules librate? *J. Phys. Chem.* 96: 10842–10848.

[14] Cho, M., Du, M., Scherer, N.F., Fleming, G.R. and Mukamel, S. (1993). Off-resonance birefringence in liquids. *J. Chem. Phys.* 99: 2410–2428.

[15] Righini, R. (1993). Ultrafast optical Kerr effects in liquids and solids. *Science* 262: 1386–1390.

Foggi, P., Bellini, M., Kien, D.P., Vercuque, I. and Righini, R. (1997). Relaxation dynamics of water and HCl aqueous solutions measured by time-resolved optical Kerr effect. *J. Phys. Chem.* 101: 7029–7035.

Ricci, M., Bartolini, P., Chelli, R., Cardini, G., Califano, S. and Righini, R. (2001). The fast dynamics of benzene in the liquid phase, Part I: Optical Kerr effect experimental investigation. *Phys. Chem. Chem. Phys.* 3: 2795–2802.

[16] Chang, Y.J. and Castner Jr., E.W. (1993). Femtosecond dynamics of hydrogen-bonding solvents. Formamide and N-methylformamide in acetonitrile, DMF, and water. *J. Chem. Phys.* 99: 113–125.

Chang, Y.J. and Castner Jr., E.W. (1993). Fast responses from "slowly relaxing" liquids: A comparative study of the femtosecond dynamics of triacetin, ethylene glycol, and water. *J. Chem. Phys.* 99: 7289–7299.

Castner Jr., E.W., Chang, Y.J., Chu, Y.C. and Walrafen, G.E. (1995). The intermolecular dynamics of liquid water. *J. Chem. Phys.* 102: 653–659.

Chang, Y.J. and Castner Jr., E.W. (1996). Intermolecular dynamics of substituted benzene and cyclohexane liquids, studied by femtosecond nonlinear-optical polarization spectroscopy. *J. Phys. Chem.* 100: 3330–3343.

[17] Deuel, H.P., Cong, P. and Simon, J.D. (1994). Probing intermolecular dynamics in liquids by femtosecond optical Kerr effect Spectroscopy: Effects of molecular symmetry. *J. Phys. Chem.* 98: 12600–12608.

Cong, P., Deuel, H.P. and Simon, J.D. (1995). Structure and dynamics of molecular liquids investigated by optical-heterodyne detected Raman-induced Kerr effect spectroscopy (OHD-RIKES). *Chem. Phys. Lett.* 240: 72–78.

Chang, Y.J., Cong, P. and Simon, J.D. (1995). Optical heterodyne detection of impulsive stimulated Raman scattering in liquids. *J. Phys. Chem.* 99: 7857–7859.

Cong, P., Chang, Y.J. and Simon, J.D. (1996). Complete determination of intermolecular spectral densities of liquids using position-sensitive Kerr lens spectroscopy. *J. Phys. Chem.* 100: 8613–8616.

[18] Quivetis, E.L. and Neelakandam, M. (1996). Femtosecond optical Kerr effect studies of liquid methyl iodide. *J. Phys. Chem.* 100: 10005–10014.

[19] Bartolini, P., Ricci, M., Torre, R. and Righini, R. (1999). Diffusive and oscillatory dynamics of liquid iodobenzene measured by femtosecond optical Kerr effect. *J. Chem. Phys.* 110: 8653–8659.

[20] Loughnane, B., Scodinu, A., Farrer, R.A. Fourkas, J.T. and Mohanty, U. (1999). Exponential intermolecular dynamics in optical Kerr effect spectroscopy of small-molecule liquids. *J. Chem. Phys.* 111: 2686–2694.

Loughnane, B., Scodinu, A. and Fourkas, J.T. (2006). Temperature-dependent optical Kerr effect spectroscopy of aromatic liquids. *J. Phys. Chem. B* 110: 5708–5720.

[21] Voehringer, P. and Scherer, N.F. (1995). Transient grating optical heterodyne detected impulsive stimulated Raman scattering in simple liquids. *J. Phys. Chem.* 99: 2684–2695.

Winkler, K., Lindner, J., Buoersing, H. and Voehringer, P. (2000). Ultrafast Raman-induced Kerr-effect of water: Single molecule versus collective motions. *J. Chem. Phys.* 113:4674–4682.

Winkler, K., Lindner, J. and Voehringer, P. (2002). Low frequency depolarized Raman-spectral density of liquid water from femtosecond optical Kerr-effect measurements: Lineshape analysis of restricted translational modes. *Phys. Chem. Chem. Phys.* 4: 2144–2155.

[22] Torre, R., Santa, I. and Righini, R. (1993). Pre-transitional effects in the liquid-plastic transition of *p*-terphenyl. *Chem. Phys. Lett.* 212:90–95.

Torre, R. and Califano, S.(1996). Local order effect on molecular orientational dynamics: Time resolved non-linear spectroscopy. *J. Chim. Phys.* 93:1843–1857.

[23] Stankus, J.J., Torre, R. and Fayer, M.D. (1993). Influence of local liquid structure on orientational dynamics: Isotropic phase of liquid crystals. *J. Chem. Phys.* 97: 9480–9487.

Torre, R., Ricci, M., Saielli, G., Bartolini, P. and Righini, R. (1995). Orientational Dynamics in the isotropic phase of a nematic mixture: Sub-picosecond time resolved optical Kerr effect experiments on ZLI-1167 liquid crystal. *Mol. Cryst. Liq. Cryst.* 262: 391–402.

Torre, R., Tempestini, F., Bartolini, P. and Righini, R. (1995). Collective and single particle dynamics near the isotropic-nematic phase transition. *Philos. Mag. B* 77: 645–653.

[24] Torre, R., Bartolini, P. and Pick, R.M. (1998). Time-resolved optical Kerr effect in a fragile glass-forming liquids, salol. *Phys. Rev. E* 57: 1912–1920.

[25] Torre, R., Bartolini, P., Ricci, M. and Pick, R.M. (2000). Time-resolved optical Kerr effect on a fragile glass-forming liquid: Test of different mode-coupling theory aspects. *EuroPhys. Lett.* 52: 324–329.

[26] Hinze, G., Brace, D.D., Gottke, S.D. and Fayer, M.D. (2000). A detailed test of mode-coupling theory on all time scales: Time domain studies of structural relaxation in a supercooled liquid. *J. Chem. Phys.* 113: 3723–3733.

[27] Ricci, M., Bartolini, P. and Torre, R. (2002). Fast dynamics of a fragile glass-former by time resolved spectroscopy. *Philos. Mag. B* 82: 541–551.

[28] Prevosto, D., Bartolini, P., Torre, R., Ricci, M., Taschin, A., Capaccioli, S., Lucchesi, M. and Rolla, P. (2002). Relaxation processes in a epoxy resin studied by time resolved optical Kerr effect. *Phys. Rev. E* 66: 011502 (1–12).

[29] Pratesi, G., Bartolini, P., Senatra, D., Ricci, M., Righini, R., Barocchi, F. and Torre, R. (2003). Experimental studies of the ortho-toluidine glass transition. *Phys. Rev. E* 67: 021505(1–8).

[30] Cang, H., Novikov, V.N. and Fayer, M.D. (2003). Experimental observation of a nearly logarithmic decay of the orientational correlation function in supercooled liquids on the picosecond-to-nanosecond time scales. *Phys. Rev. Lett.* 90: 197401(1–4).

[31] Hinze, G., Francis, R. and Fayer, M.D. (1999). Translational–rotational coupling in supercooled liquids: Heterodyne detected density induced molecular alignment. *J. Chem. Phys.* 111:2710–2719.

[32] Elschner, R., Macdonald, R., Eichler, H.J., Hess, S. and Sonnet, A.M. (1999). Molecular reorientation of a nematic glass by laser-induced heat flow. *Phys. Rev. E* 60: 1792–1797.

[33] Torre, R., Taschin, A. and Sampoli, M. (2001). Acoustic and relaxation processes in supercooled o-ter-phenyl by optical-heterodyne transient grating experiment. *Phys. Rev. E* 64: 61504(1-10).

[34] Taschin, A., Torre, R., Ricci, M., Sampoli, M., Dreyfus, C., Pick. and R.M. (2001). Translational-rotation coupling in transient grating experiments: Theoretical and experimental evidences. *EuroPhys. Lett* 53: 407–413.

[35] Glourieux, C., Hinze, G., Nelson, K.A. and Fayer, M.D. (2002). Thermal, structural, and orientational relaxation of supercooled salol studied by polarization-dependent impulsive stimulated scattering. *J. Chem. Phys.* 116: 3384–3395.

[36] Ricci, M., Wiebel, S., Bartolini, P., Taschin, A. and Torre, R. (2004). Time-resolved optical Kerr effect experiments on supercooled benzene and test of mode-coupling theory. *Philos. Mag.* 84: 1491–1498.

[37] Torre, R., Bartolini, P. and Righini, R. (2004). Structural relaxation in supercooled water by time-resolved spectroscopy. *Nature* 428: 296–299.

[38] Constantine, S., Zhou, Y., Morais, J. and Zigler, L.D. (1997). Dispersed optical heterodyne detected birefringence and dichroism of transparent liquids. *J. Phys. Chem. A* 101: 5456–5462.

[39] Bloembergen, N. (1965). *Nonlinear optics*. W.A. Benjamin, New York.

[40] Shen, Y.R. (1984). *The principles of nonlinear optics*. Wiley, New York.

[41] Hellwarth, R.W. (1977). Third-order optical susceptibilities of liquids and solids. *Prog. Quantum Electron.* 5:1–68.

[42] Berne, B.B. and Pecora, R. (1976). *Dynamic light scattering*. Wiley, New York.

[43] Kubo, R., Toda, M. and Hashitsume, N. (1992). *Statistical Physics I and II*. Series on Solid-State Science, vol.31, Springer-Verlag, Berlin.

[44] Madden, P.A. (1991). Molecular motion in liquids. In Hansen, J.P., Levesque, D., and Zinn-Justin, J., editors, *Liquids, freezing and glass transition*, Les Houches 1989, Session LI, pages 551–624, North-Holland, Amsterdam.

[45] Latz, A. and Letz, M. (2001). On the theory of light scattering in molecular liquids. *Eur. Phys. J. B* 19: 323–343.

[46] Ladanyi, B.M. and Klein, S. (1996). Contributions of rotation and translation to polarizability anisotropy and solvation dynamics in acetonitrile. *J. Chem. Phys.* 105: 1552–1561.

[47] Murry, R.L., Fourkas, J.T. and Keyes, T. (1998). Nonresonant intermolecular spectroscopy beyond the Placzek approximation. I. Third-order spectroscopy. *J. Chem. Phys.* 109: 2814–2825.

[48] Pick, R.M., Franosch, T., Latz, A. and Dreyfus, C. (2003). Light-scattering by longitudinal phonons in molecular supercooled liquids. I. Phenomenological approach. *Eur. Phys. J. B* 31:217–229.

Franosch, T., Latz, A. and Pick, R.M. (2003). Light-scattering by longitudinal phonons in molecular supercooled liquids. II. The Microscopic Derivation of the phenomenological equations. *Eur. Phys. J. B* 31:229–246.

[49] Pick, R.M., Dreyfus, C., Azzimani, A., Gupta, R., Torre, R., Taschin, A. and Franosch, T. (2004). Heterodyne detected transient gratings in supercooled molecular liquids. A phenomenological theory. *Eur. Phys. J. B* 39:169–197.

Franosch, T. and Pick, R.M. (2005). Transient grating experiments on supercooled molecular liquids. II. Microscopic derivation of the phenomenological equations. *Eur. Phys. J. B* 47: 341–361.

[50] Gray, C.G. and Gubbins, K.E. (1984). *Theory of molecular fluids*. Clarendon Press, Oxford.

[51] Hansen, J.P. and McDonald, I.R. (1986). *Theory of simple liquids*. Academic Press, London.

[52] Balucani, U. and Zoppi, M. (1994). *Dynamics of the liquid state*. Clarendon Press, Oxford.

[53] DeBenedetti, P.G. (1996). *Metastable liquids*. Princeton University Press, New Jersey.

[54] Landau, L.D. and Lifhsitz, E.M. (1959). *Fluids mechanics*. Pergamon Press, London.

[55] Boon, J.P. and Yip, S. (1980). *Molecular hydrodynamics*. McGraw-Hill, New York.

[56] Wang, C.H. (1985). *Spectroscopy of condensed media*. Academic Press, New York.

[57] Mukamel, S. (1995). *Principles of non linear optical spectroscopy*. Oxford University Press, New York.

[58] Chelli, R., Cardini, G., Ricci, M., Bartolini, P., Righini, R. and Califano, S. (2001). The fast dynamics of benzene in the liquid phase, Part II: A molecular dynamics simulation. *Phys. Chem. Chem. Phys.* 3: 2803–2810.

[59] Ryu, S. and Stratt, R.M. (2004). A case study in the molecular interpreta-
      tion of optical Kerr effect spectra: Instantaneous-normal-mode analysis of
      the OKE spectrum of liquid benzene. *J. Phys. Chem. B* 108: 6782–6795.
[60] Götze, W. (1991). Aspects of structural glass relaxations. In Hansen, J.P.,
      Levesque, D., and Zinn-Justin, J., editors, *Liquids, freezing and glass
      transition*, Les Houches 1989, Session LI, pages 287–499, North-Holland,
      Amsterdam.

      Götze, W. and Sjögren, L. (1992). Relaxation processes in supercooled
      liquids. *Rep. Prog. Phys.* 55: 241–376.
[61] Kob, W. (1996). Theoretical perspectives on supercooled liquids. In
      Fourkas, J.T., Kivelson, D., Mohanty, U. and Nelson, K.A., editors, *Super-
      cooled Liquids*, ACS Symposium Series, p. 28–44. Princeton University
      Press, New Jersey.
[62] Cummins, H.Z. (1999). The liquid–glass transition: A mode-coupling
      perspective. *J. Phys. Condens. Matter* 11: A95–A117.
[63] Keyes, T. and Kivelson, D. (1972). Depolarized light scattering: Theory
      of the sharp and broad Rayleigh lines. *J. Chem. Phys.* 56: 1057–1065.
[64] Moro, G.J., Nordio, P.L., Noro, M. and Polimeno, A. (1994). A cage
      model of liquids supported by molecular dynamics simulations. I. The
      cage variables. *J. Chem. Phys.* 101: 693–702.

      Polimeno A. and Moro, G.J. (1994). A cage model of liquids supported
      by molecular dynamics simulations. II. The stochastic model. *J. Chem.
      Phys.* 101: 703–712.
[65] Götze, W. (1999). Recent tests of the mode-coupling theory for glassy
      dynamics. *J. Phys. Condens. Matter* 11: A1–A45.
[66] Götze, W. and Sperl, M. (2004). Nearly-logarithmic decay of correlations
      in glass-forming liquids. *Phys. Rev. Lett.* 92: 105701.
[67] L. Sjögren (1986). Diffusion of impurities in a dense fluid near the glass
      transition. *Phys. Rev. A* 33:1254–1260.

      Götze, W. and Voigtmann, Th. (2000). Universal and non-universal fea-
      tures of glassy relaxation in propylene carbonate. *Phys. Rev. E* 61:
      4133–4145.

      Chong, S.-H. and Götze, W. (2002). Structural relaxation in a system of
      dumbbell molecules. *Phys. Rev. E* 65: 051201.
[68] Sperl, M. (2006). Cole-Cole law for critical dynamics in glass-forming
      liquids. *Phys. Rev. E* 74: 011503(1–15).
[69] Dies, J.C. and Rudolph, W. (1996). *Ultrashort laser pulse phenomena.*
      Academic Press, San Diego.
[70] Bartolini, P., Eramo, R., Taschin, A., De Pas, M. and Torre, R. (2007).
      A real-time acquisition system for pump-probe spectroscopy. *Philos. Mag.*
      87:731–740.

[71] Giraud, G., Gordon, C.M., Dunkin, I.R. and Wynne, K. (2003). The effects of anion and cation substitution on the ultrafast solvent dynamics of ionic liquids: A time-resolved optical Kerr-effect spectroscopic study. *J. Chem. Phys.* 119: 464–477.

[72] Bartolini, P., Eramo, R., Taschin, A. and Torre, R. (2007). A super-heterodyne-detected optical-Kerr-effect study of supercooled water dynamics. In preparation.

[73] Lotshaw, W.T., Mc Morrow, D., Thantu, N., Melonger, J.S. and Kitchenham, R. (1995). Intermolecular vibrational coherence in molecular liquids. *J. Raman Spectrosc.* 26: 571–583.

[74] Kinoshita, S., Kai, Y., Yamaguchi, M. and Yagi, T. (1995). Direct comparison between ultra-fast optical Kerr effect and high-resolution light scattering spectroscopy. *Phys. Rev. Lett.* 75: 148–151.

[75] Brodin, A. and Rössler, A.E. (2006). Depolarized light scattering versus optical Kerr effect spectroscopy of supercooled liquids: Comparative analysis. *J. Chem. Phys.* 125: 114502(1-9).

Chapter 3

TRANSIENT GRATING EXPERIMENTS IN GLASS-FORMER LIQUIDS

A Unique Tool to Investigate Complex Dynamics

Andrea Taschin, Roberto Eramo, Paolo Bartolini, and Renato Torre

Abstract   The transient grating experiment is a very useful spectroscopic tool to investigate glass-former and viscous liquid dynamics. The technical improvements, introduced in the recent years, transformed it in a unique spectroscopic technique, thanks to the wide time window covered and the quality of experimental data. Furthermore, only recently a precise definition of the measured response function has been obtained permitting a more exact description of several dynamical processes present in complex liquids. In this chapter we review the transient grating spectroscopy from both experimental and theoretical points of view, paying particular attention to the interpretation of the measured data.

3.1   Introduction

The transient grating (TG) technique was known since the end of the 1970s, when spectroscopists discovered that it was possible to induce within a material a spatial periodic modulation of the dielectric constant using the interference field produced by a pair of laser pulses [1]. They realized also that, by monitoring the relaxation toward equilibrium of this induced transient grating by means of diffraction of another beam, it was possible to infer important information on the dynamic proprieties of the material [2,3]. This optical spectroscopic technique is based on nonlinear optical effects [4], and it is indeed a particular case of the large category of the so called four-wave-mixing techniques [5–7]. The TG technique has been shown to be a very useful and flexible method to investigate the dynamics of isotropic materials. Depending on the characteristics of the material, on the laser radiation and its interaction with the sample, the TG experiments enable the selective measurement of many static and dynamical proprieties of the investigated matter. In particular, transient grating techniques turn out to be a powerful means for the study of supercooled liquids and glasses, providing an unique experimental insight in the relaxation dynamics of these materials.

The first studies on glassy liquids by means of transient grating experiments date back to the end of the 1980s and the beginning of the 1990s [1, 8–10]. In those years new appellations for transient grating experiments were also coined. Since in these experiments the two pulses excite coherently some material modes, differently from the Light Scattering (LS) experiment where these modes are thermally activated, they have been also called *impulsive stimulated light scattering* (ISS) experiments [11]. Moreover, depending on the frequency of the excited modes, they have been named: *impulsive stimulated Brillouin scattering* (ISBS) experiments when acoustic modes are excited, *impulsed stimulated Raman scattering* (ISRS) experiments when optical phonons or molecular vibrational modes are excited and, finally, *impulsive stimulated thermal scattering* (ISTS) experiments when a thermal grating is build up by weak absorption of pumps. This last kind of experiments has been shown to be a powerful means for the study of the structural dynamics in supercooled liquids.

The transient grating experiments were initially performed with pulsed probe. In 1995, Yang and Nelson [12] introduced, for first time, a ISTS experiment with a continuous-wave probe, where the whole temporal response was recorded in a single laser shot. This improvement enabled to increase greatly the signal to noise ratio of the data and widely enlarge the temporal window of the experiment (from nsec to msec). This last feature turned out to be of fundamental importance to cover wide range of scale times over which the glass relaxation dynamics occurs.

Another important improvement of TG experiments has been obtained in 1998 [13, 14] with the introduction of a heterodyne detection setup using a diffractive optical element. The new configuration simplified greatly the optical setup and provided furthermore a very stable phase locking between probe and reference beams, feature of fundamental relevance to realize an heterodyne detection.

In reference to the interpretation of the data, recently a series of heterodyne detected TG (HD-TG) experiments on supercooled liquids have shown that several approximations introduced in the hydrodynamic model must be reconsidered to fully explain the measured signal. HD-TG experiments on *ortho*-terphenyl, glycerol and 3-methylpentane [15–18] have shown that the measured signal is sensitive to the interaction between temperature and density dynamics. Other works on water [19–22] underline the importance of considering the dielectric constant change induced directly by the temperature, generally neglected in the interpretation of the transient grating experiment data. Other HD-TG experiments [23–28], with the polarization control of the laser fields and performed on supercooled anisotropic liquids (salol and m-toluidine), have shown the presence of a contribution in the signal that must be ascribed to the anisotropic part of the induced dielectric tensor. Other time-resolved optical experiments pointed out the importance of introducing the molecular orientational effects and the coupling with the density in the study of anisotropic supercooled liquids [29]. Hence the orientational variables cannot be neglected in the definition of the response and a complete set of hydrodynamic equations [30, 31], including temperature, density, and local orientational distributions must be considered in order to describe the measured dynamics in these viscoelastic liquids.

Definitely, the HD-TG experiments turned out to be an important breakthrough because of their capability of measuring collective relaxation times in a time window where alternative methods fail.

The present work is divided in three parts. It begins (Sect. 3.2) introducing the theoretical background needed to understand the TG experiments and to define the measured TG signals. Here we report, at first, a simplified description of TG experiments then a theoretical definition of the experiment into the framework of the nonlinear optics, in particular of the four-wave-mixing theory. Finally, we will analyze the TG signal using several hydrodynamic models. In Sect. 3.3, we report a detailed account of the laser system and optical set-up used for our experiment and the description of the heterodyne detection, underlining the important experimental improvements achieved by using this technique. The last section (Sect. 3.4) is devoted to an overview of the experimental results obtained on different liquids.

3.2 Theory of the Transient Grating Experiment

In a transient grating experiment, two high power laser pulses, obtained by dividing a single laser beam, interfere within the sample to produce a spatially periodic variation of the optical properties of the material by standing-wave excitation of some material mode [1, 7, 11]. This modulation can be probed by a third laser beam, typically of different wavelength that can be a pulsed or a continuous-wave (CW) field. It impinges on the induced grating and is subsequently diffracted by it. A measurement of the diffracted intensity permits the acquisition of dynamic information on the relaxing TG and, consequently of the properties of the excited modes. A schematic drawing of a TG experiment is shown in Fig. 3.1.

The spatial modulation is characterized by the wave-vector $\mathbf{q}$, given by the fields wave-vector difference $(\mathbf{k}_1 - \mathbf{k}_2)$ of the two pumps. Its modulus is

$$q = \frac{4\pi \, \sin (\theta_{ex})}{\lambda_{ex}} \, , \tag{3.1}$$

where λ_{ex} and θ_{ex} are the wavelength and the incidence angle of the exciting pumps, respectively. If the depth of the transient grating is much larger than

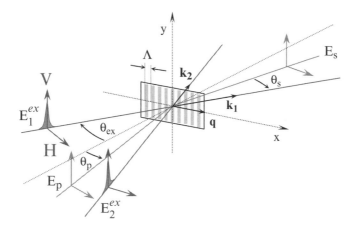

Fig. 3.1 Transient grating experiment. Two excitation pulses, E_1^{ex} and E_2^{ex} induce a standing-wave material response with period Λ, whose relaxation is monitored by Bragg scattering of a third beam, E_p. The possible polarization directions for each beam are also reported

grating spacing, it can be probed efficiently only if the wave-vectors of probe and scattered beams obey the *Bragg* condition

$$\mathbf{k_s} - \mathbf{k_p} = \mathbf{q}m, \quad m = \pm 1, \pm 2, \dots \tag{3.2}$$

If the frequencies of excited modes are very low with respect to the optical frequency, we have $|\mathbf{k_s}| \simeq |\mathbf{k_p}|$. Thus the λ_p value and (3.2) fix completely the directions of the probe and the diffracted beams:

$$\mathbf{k_s} - \mathbf{k_p} = \pm \mathbf{q} . \tag{3.3}$$

The modification of the optical properties of the medium can occur in the real part of the refractive index (birefringence and/or phase grating) or in its imaginary part (dichroic and/or amplitude grating) [1]. When the pumping fields are not resonant with any material levels, the induced grating is mainly due to the birefringence-phase grating. We will focus our attention on TG experiments where we have a weak absorption of pumping beams but where, thanks to the fast thermalization of the absorbed energy, the induced grating is once again only dominated by the birefringence-phase contribution.

Let us analyze the main effects responsible for the birefringence-phase grating in this kind of TG experiment. The absorbed energy quickly thermalizes, yielding a temperature grating. Thanks to the thermal expansion, a density and a pressure grating build up. The pressure grating launches two counter-propagating acoustic waves whose superposition makes a standing wave. This wave oscillates at a period dependent on the material sound velocity and on the induced q-value, and decays in time with an exponential law. The thermal grating decays by heat diffusion following again an exponential law but with times generally much longer than the period and damping of acoustic oscillations.

A density grating can build up also without absorption by the electrostrictive force. The interference field induces an electric dipole in every molecule and the induced dipole moves toward the areas of greatest field intensity. Unlike the thermal grating, now we have compression zones in the maximum light intensity points. This grating produces the same acoustic standing wave but with a phase difference of $\pi/2$ with respect to the previous situation.

Both thermal and electrostriction effects induce a molecular velocity grating or a strain grating. In liquid formed of anisotropic molecules this strain grating tends to align the molecules by means of the *rotation–translation coupling* (R–TC) effect [23, 24, 26, 28, 29, 32–35] giving rise to a birefringence grating. In this case, the diffracted signal depends on the probe polarization. A birefringence grating can also be built directly by the pump electric field (hereafter named pump-induced birefringence, PIB). In anisotropic molecules, in fact, the electric dipole is induced mainly in a molecular direction; the same field, then, applies a torque to the dipoles, and partly orients them, producing again a birefringence grating. Moreover, by means of the same R–TC effect, this grating generates either transverse or longitudinal phonons for pumps with perpendicular or parallel polarization respectively. Clearly, the diffracted signal now depends also on the polarization of the two pumps.

So the exciting fields, indeed, produce three excitation forcing terms that drive the material modes: a heat deposition that drives directly the temperature, T, an electrostriction that drives the velocity, $\dot{\rho}$, and an electric-torque that drives the local orientational distribution of the molecules, Q_{ij}.

All of these forcing terms are normally present but, depending on the radiation–material interaction nature, some of them may not be effective. The excitation modifies the optical properties of the material, namely the dielectric constant, whose relaxation is directly connected with the relaxation functions of the material modes, ρ, T, and Q_{ij}. Which mode is really effective in defining the dielectric relaxation depend on which one of the forcing components is active and on the equations governing the mode dynamics. In other words, since the equations that define the dynamics of thermal, density, and orientational variables are coupled, more then one material mode has to be taken into account, even if a single forcing term is active. For example [23–27] in m-toluidine glass-former the heating effect is the main driving term but, because of R–TC effects, both the density and the local orientation variables are present in the relaxation process. Moreover, in supercooled water [19–22] it has been shown how the dielectric constant change, directly induced by the temperature, usually neglected either in the light scattering or in TG interpretation, must be taken into account. Indeed only a careful analysis of the experimental results will help

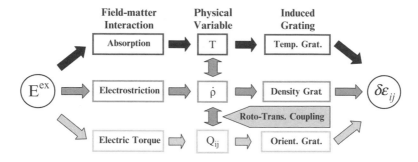

Fig. 3.2   The excitation forces and the induced modes in a TG experiment. The possible couplings among modes are also reported

to understand which forcing term and material mode is active. In Fig. 3.2 we report a sketch of the all possible excitation channels and the different involved modes in a TG experiment. The figure shows also how the the induced modes can be generally coupled.

3.2.1   TG and Four-Wave Mixing

In the previous section, we introduced a simplified description of a TG experiment, where the excitation and probing were considered as separate processes. In fact, diffraction by transient grating belongs to a more general context of light–matter interaction. Different approaches to the problem exist with varying levels of complexity according to the physical processes involved and to the material properties studied.

A more rigorous approach is to consider the problem in the framework of nonlinear optics, in particular in the four-wave mixing (FWM) context [6, 7]. Here the TG diffraction is explained as optical mixing of the three incident fields, the two pumps and the probe, which, by interaction with the medium, give a nonlinear polarization that becomes the source of the diffracted field. This approach gives, moreover, a rigorous proof of the separation of the exciting and probing processes.

3.2.1.1   The Induced Third Order Polarization and Excitation Process

Here we report the result of the calculation of the third-order polarization based on a model of the nonlinear radiation–matter interaction introduced by Hellwarth in 1977 [5]. This is a semiclassical model that allows us to define theoretically the nonlinear polarization in condensed matter using a perturbative approach. Thus following the above approach, the third-order polarization $P_i^{(3)}(\mathbf{r}, t)$ at the point $\mathbf{r}$ and at the time t for homogenous media can be written as spatial and temporal convolution of the electric field with a response function[1]:

$$P_i^{(3)}(\mathbf{r}, t) = \int d\mathbf{r}_1 \, d\mathbf{r}_2 \, d\mathbf{r}_3 \int dt_1 \, dt_2 \, dt_3$$

$$\mathcal{R}_{ijkl}^{(3)}(\mathbf{r} - \mathbf{r}_1, \mathbf{r} - \mathbf{r}_2, \mathbf{r} - \mathbf{r}_3, t - t_1, t - t_2, t - t_3)$$

$$\times E_j(\mathbf{r}_1, t_1) \, E_k(\mathbf{r}_2, t_2) \, E_l(\mathbf{r}_3, t_3). \quad (3.4)$$

E_i is the total classical electric field, which is the sum of the three incoming fields (the two pumps and the probe) and the indices i, j, k, l refer to the x, y, z component of $\mathbf{P}^{(3)}$ and $\mathbf{E}$. Here and in the following repeated space indices are assumed to be summed. The function $\mathcal{R}_{ijkl}^{(3)}$ is the *third-order response function*, or *third-order susceptibility* in the frequency domain: it contains the complete microscopic information about the system and hence about all its equilibrium dynamic properties.

By assuming that the incident electromagnetic frequencies are much lower than any electronic gap frequencies, it is possible to apply the Born-Oppenheimer (BO) separation and, as already reported in Chap. 2 (2.14), the expression of the third-order polarization can be written as the sum of an electronic contribution and a nuclear contribution. The electronic part, or electronic hyper-polarizability, it is a function of the nuclear positions and it gives an account of instantaneous local response of the medium arising from nonlinear distortion of electronic clouds. This response is temperature independent at fixed density. The second term, instead, describes the polarization due to slight changes in the nuclear configurations induced by electronic distortions. The nuclear response depends on the nuclear motion relaxation time and therefore it is usually strongly temperature dependent.

If the frequencies of fields are also well above any nuclear resonance frequency, the polarization due to the nuclear response assumes the following expression [5]:

$$P_i^{(3)} = E_j(\mathbf{r}, t) \int d\mathbf{r}' \int dt' \, \mathcal{R}_{ijkl}^{(\text{nucl})}(\mathbf{r} - \mathbf{r}', t - t') \, E_k(\mathbf{r}', t') E_l(\mathbf{r}', t'). \quad (3.5)$$

Equation (3.5) suggests the separation between the excitation and the probing processes. In fact, if the probe beam is very weak with respect to the pump intensity, we can neglect the probe field in $E_k(\mathbf{r}', t')$ and $E_l(\mathbf{r}', t')$. Moreover, if the excitation pulses are very short as compared with relaxation times and we are observing times t immediately after the excitation, we can consider only the probe field in the $E_j(\mathbf{r}, t)$ field outside the integral:

$$P_i^{(3)}(\mathbf{r}, t) = E_j^{\text{p}}(\mathbf{r}, t)$$

$$\int d\mathbf{r}' \int dt' \, \mathcal{R}_{ijkl}^{(\text{nucl})}(\mathbf{r} - \mathbf{r}', t - t') \, E_k^{\text{ex}}(\mathbf{r}', t') E_l^{\text{ex}}(\mathbf{r}', t'), \quad (3.6)$$

where $E_i^{\text{ex}}(\mathbf{r}',t')$ is the field sum of the two pump fields and $E_j^{\text{p}}(\mathbf{r},t)$ the probe one. This expression can be rewritten as[2]

$$P_i^{(3)}(\mathbf{r},t) = E_j^{\text{p}}(\mathbf{r},t)\frac{1}{4\pi}\delta\epsilon_{ij}(\mathbf{r},t), \tag{3.7}$$

with

$$\delta\epsilon_{ij}(\mathbf{r},t') = 4\pi \int d\mathbf{r}' \int dt' \mathcal{R}_{ijkl}^{(\text{nucl})}(\mathbf{r}-\mathbf{r}',t-t') E_k^{\text{ex}}(\mathbf{r}',t')E_1^{\text{ex}}(\mathbf{r}',t'). \tag{3.8}$$

Thus, the excitation pulses produce, through the third-order susceptibility, a linear dielectric constant change $\delta\epsilon_{ij}$, the probe field linearly interacts with the medium giving a polarization $P^{(3)}$ which, then, is the source of the scattered field.

As we shall see in the next subsections, in a TG experiment the interesting quantity is a $\mathbf{q}$-component of the spatial Fourier transform of $\delta\epsilon_{ij}$, therefore we report, in that follows, (3.8) in the $\mathbf{q}$-space:

$$\delta\epsilon_{ij}(\mathbf{q},t) = 4\pi \int dt' \ \mathcal{R}_{ijkl}^{(\text{nucl})}(\mathbf{q},t-t') \ F_{kl}(\mathbf{q},t'), \tag{3.9}$$

with $F_{kl}(\mathbf{q},t)$ spatial Fourier transform of

$$F_{kl}(\mathbf{r}',t') = E_k^{\text{ex}}(\mathbf{r}',t')E_1^{\text{ex}}(\mathbf{r}',t'). \tag{3.10}$$

Let us suppose that the spot sizes of the excitation pulses are much larger than the induced grating spacing. Then we can approximate the excitation fields as plane waves. According to this approximation, the excitation fields can be expressed as

$$\mathbf{E}_n^{\text{ex}}(\mathbf{r},t) = \hat{\mathbf{e}}_n E_{\text{ex}}(t) \exp\left[i\left(\mathbf{k}_n \cdot \mathbf{r} - \omega_{\text{ex}}t\right)\right] + \text{c.c.} ; \quad n = 1, 2, \tag{3.11}$$

where $E_{\text{ex}}(t)$ is the pulse time profile. Here we assume that the two pulses have the same frequency and intensity and are time coincident. Moreover, if we suppose the excitation pulses to be much shorter than the oscillation period of the excited material modes, we can consider the induced grating to be instantaneous. The introduced approximations, lead to the condition of the *ideal excitation state* for a transient grating experiment. According to (3.11) the excitation term (3.10) becomes

$$F_{kl}(\mathbf{r},t) = 2E_{\text{ex}}^2(t)\left[(\hat{\mathbf{e}}_{1k}\hat{\mathbf{e}}_{1l} + \hat{\mathbf{e}}_{2k}\hat{\mathbf{e}}_{2l}) + \cos(\mathbf{q}_0 \cdot \mathbf{r}) U_{kl}\right], \tag{3.12}$$

where $\mathbf{q}_0 = \mathbf{k}_1 - \mathbf{k}_2$ and we have defined the tensor $U_{kl} = (\hat{\mathbf{e}}_{1k}\hat{\mathbf{e}}_{2l} + \hat{\mathbf{e}}_{2k}\hat{\mathbf{e}}_{1l})$. Here we neglected all the terms oscillating at $2\omega_{\text{ex}}$ frequency, which, anyway, should vanish with the average on optical cycle. Transforming (3.12) in q-space we get

$$
F_{kl}(\mathbf{q},t) = (2\pi)^3\, \mathcal{E}_{ex}^2 \delta(t) \{2\delta(\mathbf{q})\, (\widehat{e}_{1k}\widehat{e}_{1l} + \widehat{e}_{2k}\widehat{e}_{2l})
$$
$$
+ [\delta(\mathbf{q}-\mathbf{q}_0) + \delta(\mathbf{q}+\mathbf{q}_0)]\, U_{kl}\}, \quad (3.13)
$$

where we substituted $E_{ex}^2(t) = \mathcal{E}_{ex}^2 \delta(t)$ with $\mathcal{E}_{ex}^2$ proportional to the pump field energy. The $\delta(\mathbf{q})$ term represents a spatially uniform excitation that does not contribute to the induced grating, which is instead described by the two remaining terms.

3.2.1.2 Generated Field and Probe Process

Now we want to express the new generated beam produced by the induced $P^{(3)}$, this corresponds to the scattered field. As a first approximation in our experiment, we can partially de-couple the equations that describe the FWM. We can consider that the pumps and probe produce the $P^{(3)}$ term and we can assume their propagation is unaffected by the latter, and that the diffracted beam is entirely generated by the $P^{(3)}$ term acting as a source in the wave equation [6]. According to this approximation, the excitation and probe beams propagate in the medium with negligible modification and the scattered field $\mathbf{E}_s$ (for nonmagnetic, nonconducting, isotropic, and homogeneous media) is defined by the following equation

$$
\nabla \wedge [\nabla \wedge \mathbf{E}_s(\mathbf{r},t)] + \frac{\epsilon}{c^2}\frac{\partial^2}{\partial t^2}\mathbf{E}_s(\mathbf{r},t) = -\frac{4\pi}{c^2}\frac{\partial^2}{\partial t^2}P^{(3)}(\mathbf{r},t), \quad (3.14)
$$

where c is the speed of light in vacuum and ϵ the unperturbed dielectric constant. Using the Green's function approach and (3.8) as an expression of $P^{(3)}$, we obtain the following result for the scattered field

$$
\mathbf{E}_s(\mathbf{r},t) \propto -\frac{1}{c^2}\int_{-\infty}^{\infty} dt'd^3r' \frac{\mathbf{I}}{|\mathbf{r}-\mathbf{r}'|}\delta\left(t - t' - \frac{|\mathbf{r}-\mathbf{r}'|}{c_m}\right)
$$
$$
\cdot \frac{\partial^2}{\partial t'^2}[\delta\epsilon(\mathbf{r}',t') \cdot \mathbf{E}_p(\mathbf{r}',t')], \quad (3.15)
$$

with $c_m = c/\sqrt{\epsilon}$ the speed of light in the medium. Let us consider the probe as CW field

$$
\mathbf{E}_p(\mathbf{r},t) = \mathbf{E}_p'(\mathbf{r})\exp[i(\mathbf{k}_p\cdot\mathbf{r} - \omega_p t)] + c.c. \quad (3.16)
$$

Introducing the slowly varying amplitude approximation for $\delta\epsilon$ and performing the integral in t', (3.15) becomes

$$
\mathbf{E}_s(\mathbf{r},t) \propto \frac{\omega_p^2}{c^2}\int_{-\infty}^{\infty} d^3r' \frac{1}{|\mathbf{r}-\mathbf{r}'|}\delta\epsilon\left(\mathbf{r}',t - \frac{|\mathbf{r}-\mathbf{r}'|}{c_m}\right) \cdot \mathbf{E}_p'(\mathbf{r}')
$$
$$
\times \exp\left[i\left(\mathbf{k}_p\cdot\mathbf{r}' - \omega_p\left(t - \frac{|\mathbf{r}-\mathbf{r}'|}{c_m}\right)\right)\right] + c.c. \quad (3.17)
$$

If we make the approximation that the size of sample is small compared to the distance r and choose the coordinate system in such a way that the scattering volume is centered at $\mathbf{r}_0 = 0$, we can approximate $|\mathbf{r} - \mathbf{r}'| \simeq |\mathbf{r}| = r$ in the denominator of (3.17) and $|\mathbf{r} - \mathbf{r}'| \simeq r - \hat{\mathbf{r}} \cdot \mathbf{r}'$ in the argument of exponential of the same equation. Knowing that $\omega_p/c_m = k_p$ and supposing that the frequency change between probe and diffracted beams is negligible, $k_p \simeq k_s$, (3.17) becomes

$$\mathbf{E}_s\left(\mathbf{r}, t\right) \propto \frac{\omega_p^2}{c^2 r} \int d^3 r' \delta\epsilon\left(\mathbf{r}', t_p\right) \exp\left(-i\omega_p t_p - i\mathbf{q} \cdot \mathbf{r}'\right) \cdot \mathbf{E}_p'\left(\mathbf{r}'\right) + \text{c.c.},$$

$$(3.18)$$

where $\mathbf{q} = \mathbf{k}_s - \mathbf{k}_p$, $\mathbf{k}_s = k_p \hat{\mathbf{r}}$, and $t_p = t - r/c_m$ is the time t subtracted by the time spent by the diffracted beam to go from the sample to the $\mathbf{r}$ point. The single time t_p dependence of the integral argument means that one neglects the sample crossing time of the probe field with respect to the time scale of material motion.

Finally, if we consider the spot size of probe much larger than $2\pi/q$, the integral is the spatial Fourier transform of $\delta\epsilon(\mathbf{r}, t)$

$$\mathbf{E}_s\left(\mathbf{r}, t\right) \propto \frac{\omega_p^2}{c^2 r} \delta\epsilon\left(\mathbf{q}, t_p\right) \cdot \mathbf{E}_p' \exp\left[i\left(\mathbf{k}_s \cdot \mathbf{r} - \omega_p t\right)\right] + \text{c.c.} \qquad (3.19)$$

This result shows how in a TG experiment, with the introduced approximations, the scattered field has the same probe frequency ω_p and an amplitude directly proportional to the dielectric tensor change $\delta\epsilon\left(\mathbf{q}, t_p\right)$ and thus to the response function $\mathcal{R}$ through (3.9).

3.2.1.3   The Measured Signal

In the previous subsection we calculated an expression for the diffracted beam, $\mathbf{E}_s$, connecting it to the material response function, (see (3.19) and (3.9)). We want now to introduce the effective measured signal in a TG experiment using some the introduced approximations for the pump and probe fields. Introducing the forcing term expression, reported in (3.13), into (3.9), we obtain, neglecting the $\delta(\mathbf{q})$,

$$\delta\epsilon\left(\mathbf{q}, t\right) \propto \mathcal{R}\left(\mathbf{q}_0, t\right) \left[\delta\left(\mathbf{q} - \mathbf{q}_0\right) + \delta\left(\mathbf{q} + \mathbf{q}_0\right)\right] \cdot \mathbf{U}, \qquad (3.20)$$

where $\mathcal{R}$ means the response function tensor and $\mathbf{U}$ is the tensor defined in (3.12). Finally the scattered field becomes

$$\mathbf{E}_s\left(\mathbf{r}, t\right) \propto \frac{\omega_p^2 \mathcal{E}_{ex}^2}{c^2 r} \mathcal{R}\left(\mathbf{q}_0, t_p\right) \left[\delta\left(\mathbf{q} - \mathbf{q}_0\right) + \delta\left(\mathbf{q} + \mathbf{q}_0\right)\right]$$
$$\cdot \mathbf{U} \cdot \mathbf{E}_p' \times \exp\left[i\left(\mathbf{k}_s \cdot \mathbf{r} - \omega_p t\right)\right] + \text{c.c.} \quad (3.21)$$

The two $\mathbf{q}$-space δ functions express the Bragg condition for a thick grating delineated in Sect. 3.2 and they fix unambiguously the propagating direction of $\mathbf{E}_s$: $\mathbf{k}_s = \pm\mathbf{q}_0 + \mathbf{k}_p$. Assuming that the probe beam is a CW field incident at the Bragg angle and omitting the delay temporal shift, we get

$$\mathbf{E}_s\left(\mathbf{q},t\right) \propto \frac{\omega_p^2 \mathcal{E}_{ex}^2 \mathcal{E}_p}{c^2 r}\left[\boldsymbol{\mathcal{R}}\left(\mathbf{q},t\right)\cdot\mathbf{U}\cdot\widehat{\mathbf{e}}_p\right]\times\exp\left[\mathrm{i}\left(\mathbf{k}_s\cdot\mathbf{r}-\omega_p t\right)\right]+\text{c.c.}, \quad (3.22)$$

where $\mathbf{E}_p'=\mathcal{E}_p\widehat{\mathbf{e}}_p$; thus the scattered field is directly proportional to a projection of the response function, $\boldsymbol{\mathcal{R}}$.

With heterodyne detection (HD), we measure (see (3.62))

$$S^{\mathrm{HD}}\left(\mathbf{q},t\right)=4\left\langle\mathbf{E}_{\mathrm{loc}}\cdot\mathbf{E}_s\right\rangle_{\mathrm{op.c.}}\propto\mathcal{E}_{ex}^2\mathcal{E}_p\mathcal{E}_{\mathrm{loc}}\left(\widehat{\mathbf{d}}\cdot\boldsymbol{\mathcal{R}}\left(\mathbf{q},t\right)\cdot\mathbf{U}\cdot\widehat{\mathbf{e}}_p\right), \quad (3.23)$$

where $\mathbf{E}_{\mathrm{loc}}=\mathcal{E}_{\mathrm{loc}}\widehat{\mathbf{d}}$ is the local field (this is a CW field with the same frequency ω_p of the probe and collinear to the signal field, see Sect. 3.3.2). Since in our experiment the local field is obtained by taking a part of the probe beam, $\mathcal{E}_{\mathrm{loc}}\propto\mathcal{E}_p$, we have

$$S^{\mathrm{HD}}\left(\mathbf{q},t\right)\propto\mathcal{E}_{ex}^2\mathcal{E}_p^2\mathcal{R}\left(\mathbf{q},t\right), \quad (3.24)$$

where $\mathcal{R}\left(\mathbf{q},t\right)$ is the projection of response function tensor on the pumps, probe, and detection field polarization directions. By selecting different polarizations of the fields, different projections are probed. The above expression shows the relevance of the HD by which we are able to study directly the dynamics of the response function.

In our TG experiments we are in the ideal excitation condition because the pump-pulse duration is short compared with the relaxation times, and the excited grating is so wide in the q-direction to include a large number a grating rules. Moreover, the depth grating is so large that we have only a Bragg scattering signal. For these reasons our heterodyne signal is correctly described by formula (3.24).

3.2.1.4 The Response Function

The general definition of the nuclear response function is a very complex task, especially if absorption and thermal effects take place. A very useful method to describe and calculate this response can be found in the phenomenological theory [30,31] and models [17,22,28,36] based on an hydrodynamic description of the sample. This approach has been particularly successful for the interpretation of TG results on complex liquids. According to these models, two contributions in the response function can be identified[3]:

$$\mathcal{R}_{ijkl}^{(\mathrm{nucl})}=\mathcal{R}_{ijkl}^{(\mathrm{ISBS})}+\mathcal{R}_{ijkl}^{(\mathrm{ISTS})}. \quad (3.25)$$

We recall that the notation ISBS stays for impulsive stimulated Brillouin scattering and ISTS for impulsive stimulated thermal scattering. The first contribution is always present, while the second is present only if a thermalized light absorption takes place.

The ISBS response describes all the signal contributions generated by the excitation laser interaction with the fully nonresonant part of the liquid dielectric constant (i.e., the electrostriction and electric-torque forcing terms). Indeed ISBS is the time-resolved response equivalent to the frequency-resolved light scattering susceptibility and hence it can include not only the Brillouin but also the Rayleigh-Raman dynamic spectrum. If the sample molecule has an anisotropic polarizability, the optical Kerr effect (OKE) signal, introduced in previous chapter,[4] will be present in this response. The ISTS response is that induced by laser heat deposition in the sample (i.e., by the thermalization of molecular states excited by the laser absorption). This does not have a direct equivalent in light scattering experiments and it is particular to TG.

The ISBS response verifies the fluctuation–dissipation theorem so that it can be directly connected with the correlation functions of the dielectric constant [30]:

$$\mathcal{R}_{ijkl}^{(\text{ISBS})}(\mathbf{q}, t) \propto -\frac{\theta(t)}{k_{\text{B}}T} \frac{\partial}{\partial t} \langle \delta\epsilon_{ij}(\mathbf{q}, t) \, \delta\epsilon_{kl}(-\mathbf{q}, 0) \rangle, \qquad (3.26)$$

where $\theta(t)$ is the Heaviside step function, k_{B} is the Boltzmann constant, and T is the temperature. Equation (3.26) expresses and connects the nonequilibrium response function to equilibrium properties, in particular to the correlation function of the dielectric constant fluctuations ordinarily measured in a LS experiment. Even if this relation is expected on the basis of general theoretical considerations, verifying it turns out not to be an easy task when the response function is calculated using the hydrodynamics models [30, 31].

For the ISTS response, the connection with the dielectric correlation functions is a more complex problem. Indeed it is not immediately clear that the fluctuation-dissipation theorem must be verified. Nevertheless, under appropriate approximations it has been shown that a "pseudo fluctuation-dissipation" relation holds [30, 31], even in presence of irreversible heat diffusion processes.

The response functions, both of ISBS and ISTS type, can be expressed as a sum of elementary response functions (ERF) [30], which are strictly related to the Green's functions of the hydrodynamic equations (see the next section). Each ERF describes a single excitation–propagation–detection process taking place in a TG experiment that has been sketched in Fig. 3.2. The coefficient weighting each ERF in the sum depends on the molecular properties of the sample and on the laser field characteristics (i.e., intensity, frequency, and polarization state). The number of elementary functions appearing in the signal is strongly dependent on the hydrodynamic model used to describe the sample.

Now we want to close the section talking about the symmetry rules of the response function. The presence of nonlocal effects (i.e. a value $q \neq 0$) and the possible rotational–translational coupling in the response function produce an effective reduction of the symmetry rules as compared to the case of a generic fourth rank tensor associated with an isotropic system. In particular, when a hydrodynamic model is used to calculate the TG signal, the response function becomes a kind of "system-plus-pump-fields response," and has thus, also for an isotropic system, the spatial symmetry reduced to the rotations around the grating axis [30]. As result, the fourth-order response function tensor has eight nonvanishing components and five independent ones [30]: $\mathcal{R}_{xxxx}$, $\mathcal{R}_{yyyy}$, $\mathcal{R}_{xxyy}$, $\mathcal{R}_{yyxx}$, $\mathcal{R}_{xyxy}$ (x is parallel to q-direction and y perpendicular to the scattering plane, see Fig. 3.1). As a general result in a TG experiment: $\mathcal{R}_{xxxx} \neq \mathcal{R}_{yyyy}$ and $\mathcal{R}_{yxyx} \neq \frac{1}{2}(\mathcal{R}_{xxxx} - \mathcal{R}_{yyxx})$, differently from the OKE response where such identities fully hold, see (2.15) in the previous chapter. The full symmetry, $\mathcal{R}_{xxxx} = \mathcal{R}_{yyyy}$ and $\mathcal{R}_{yxyx} = \frac{1}{2}(\mathcal{R}_{xxxx} - \mathcal{R}_{yyxx})$, can be recovered in the limit $q \to 0$.

A further insight can be gained if we consider the ISBS and ISTS responses separately. The ISTS contribution has only four nonzero tensorial components, since $\mathcal{R}_{xyxy}^{\text{ISTS}} \equiv 0$ (in fact, excitation fields with orthogonal polarizations will induce no temperature grating). Furthermore, only two elements are independent since we have $\mathcal{R}_{xxxx}^{(\text{ISTS})} = \mathcal{R}_{xxyy}^{(\text{ISTS})}$ and $\mathcal{R}_{yyxx}^{(\text{ISTS})} = \mathcal{R}_{yyyy}^{(\text{ISTS})}$. If the rotation–translation coupling does not take place, this response turns out to be completely symmetric for all values of q, $\mathcal{R}_{xxxx}^{(\text{ISTS})} = \mathcal{R}_{yyyy}^{(\text{ISTS})} = \mathcal{R}_{xxyy}^{(\text{ISTS})} = \mathcal{R}_{yyxx}^{(\text{ISTS})}$, and only one independent component exists.

The ISBS contribution has five non-zero components but only four are independent because $\mathcal{R}_{xxyy}^{(\text{ISBS})} = \mathcal{R}_{yyxx}^{(\text{ISBS})}$. Even if the R–TC does not take place, we have anyway $\mathcal{R}_{xxxx}^{(\text{ISBS})} \neq \mathcal{R}_{yyyy}^{(\text{ISBS})}$, and $\mathcal{R}_{yxyx}^{(\text{ISBS})} \neq \frac{1}{2}(\mathcal{R}_{xxxx}^{(\text{ISBS})} - \mathcal{R}_{yyxx}^{(\text{ISBS})})$. These relations will be verified only for $q \to 0$. In this case the five elements reduce to only two and the ISBS response becomes the OKE response.

In principle, each tensor component can be experimentally selected controlling the excitation, probing and detection polarizations. In particular, we will find that symmetry breaking could lead to differences between the $\delta\epsilon_{VV}(\mathbf{r}, t)$ evolution with H pump polarization, and the $\delta\epsilon_{HH}(\mathbf{r}, t)$ evolution with V pump polarization.

3.2.2 Hydrodynamic Approach to the TG Experiments

In the previous sections, we showed the TG signal is directly connected with the dielectric response function $\mathcal{R}$. This response defines and describes the observable properties of the material in a TG experiment. An explicit function for $\mathcal{R}$ can be obtained from quantum calculations [4–6]. Indeed this turns out to be practically impossible in a complex liquid and we have to appeal to a phenomenological approach [17,22,28,30,31,36]. In this approach, first we will connect the dielectric response to the material modes by means of the familiar expansion adopted in the light scattering (LS) theory [37], then we shall use several phenomenological equations of motion to calculate the dynamics of the modes. In this way, we will able to write a relatively simple expression for the measured signal in a transient grating experiment.

According to the first-order light scattering approximation, the dielectric constant is linearly coupled to the material modes. In a molecular liquid, composed of molecules with isotropic polarizability, the only material modes involved in a light scattering experiment are the thermal and density modes. The phenomenological equation that connects the density and temperature changes to the dielectric constant change is [37]

$$\delta\epsilon_{ij}(\mathbf{q}, t) = \delta_{ij}\left(\frac{\partial\epsilon}{\partial\rho}\right)_T \delta\rho(\mathbf{q}, t) + \delta_{ij}\left(\frac{\partial\epsilon}{\partial T}\right)_\rho \delta T(\mathbf{q}, t), \qquad (3.27)$$

where in a LS experiment the $\delta\rho$ and δT are the local variation from the equilibrium value of the density and temperature, driven by the spontaneous fluctuations. In a molecular liquid with anisotropic polarizability, the birefringence effects may become appreciable and, therefore, they must be considered. Thus we have to include in (3.27) the dielectric constant change due to the molecular anisotropy. It has been proposed in [32, 33] that (3.27) becomes

$$\delta\epsilon_{ij}(\mathbf{q}, t) = \delta_{ij}\left(\frac{\partial\epsilon}{\partial\rho}\right)_T \delta\rho(\mathbf{q}, t) + \delta_{ij}\left(\frac{\partial\epsilon}{\partial T}\right)_\rho \delta T(\mathbf{q}, t) + bQ_{ij}(\mathbf{q}, t),$$
$$(3.28)$$

where b is a constant and $Q_{ij}(\mathbf{q}, t)$ is the space Fourier transform of the variable $Q_{ij}(\mathbf{r}, t)$. $Q_{ij}(\mathbf{r}, t)$ is a traceless symmetrical tensor describing the mean orientation around the $\mathbf{r}$ point for true symmetric top molecules. For most liquids, $(\partial\epsilon/\partial T)_\rho \delta T(\mathbf{q}, t)$ is very low compared to the other terms and the dielectric constant change due to the temperature can be neglected[5]

$$\delta\epsilon_{ij}(\mathbf{q}, t) = \delta_{ij}\left(\frac{\partial\epsilon}{\partial\rho}\right)_T \delta\rho(\mathbf{q}, t) + bQ_{ij}(\mathbf{q}, t). \qquad (3.29)$$

In a TG experiment, the dynamic evolution of the modes $\delta\rho$, δT, and Q_{ij} is driven by the excitation electric fields that define the forcing terms acting as source in the motion equations. However, in a LS experiment the modes are driven by the spontaneous fluctuations even if the equations of motion that define the dynamic properties of these modes are identical. In the excitation process, the interference electric field interacts with the material through different physical mechanisms. Three main effects have to be considered for near-infrared pulsed excitation in weakly absorbing materials: (a) the heat deposition due the absorption typically by the overtones and/or combination of vibrational bands, (b) the electrostriction effect, and (c) the electric torque applied to the molecules by the electric field because of the presence of an anisotropic polarizability. Strictly, also the electronic hyper-polarizability effects should be taken into account; however, since they relax in a timescale much shorter than the pulse duration, they do not contribute to the response function, and will be neglected in the following. Thus the exciting fields, produce three excitation forcing terms that drive the material modes: a heat deposition term that drives the energy equation, an electrostriction term that drives the rate of the density, and finally an electric torque term driving the local orientational distribution of the molecules.

All of these forcing terms are normally present but, depending on the nature of the radiation–matter interaction, some of them may be ineffective. The modes effective in defining the dielectric relaxation depend on which of the forcing components is active and on the equations governing the dynamics of the modes. In other words, since the equations that define the dynamics of thermal, density, and orientational variables are coupled, more than one material mode has to be taken into account, even if a single forcing term is active. For example it has been shown in the references [23, 24, 28] that in *m*-toluidine and salol glass-formers, the heating effect is the main driving term but, due to the R–TC effects, both the density and the local orientation variables are present in the relaxation process. Therefore only the analysis of the experimental results will help to understand which excitation forcing terms and material modes are active.

In the next section, we introduce the hydrodynamic equations of motion for only $\delta\rho$ and δT, and solve them when the forcing terms are present. A detailed analysis of the hydrodynamic theory including the rotational dynamics can be found in [30, 31].

3.2.2.1 Hydrodynamic Models for Simple and Complex Liquids

In the present section, we want to review some hydrodynamic models useful to describe the transient grating signal. We will start from the simplest model, introducing step by step the elements necessary to account for the complex aspects of the relaxation dynamics of the glass-former liquids, such as viscoelasticity. We will not report all the mathematical steps leading to the transient grating signal, but only the starting hydrodynamic equations and the final results, giving some remarks about the merits and faults of each model.

Hydrodynamic Model. The most simple approach is to consider the liquid as a monatomic continuous fluid without any structural relaxation effect. The motion equations, describing the relaxation toward equilibrium of the temperature and density changes ($\delta\rho$ and δT), are the well-known linearized hydrodynamic equations (LHE) [37, 38]. They can be obtained starting from continuity equations for the macroscopic densities of mass, momentum, and energy. They are the same ones used to describe the light scattering (LS) data, but with the addition of two forcing terms accounting for the modifications induced by the pump laser pulses. The LHE for the small variations of the hydrodynamic variables $\rho(\mathbf{r}, t)$, $T(\mathbf{r}, t)$, and the divergence of the fluid velocity $\psi(\mathbf{r}, t) = \nabla \cdot \mathbf{v}(\mathbf{r}, t)$ are

$$\delta\dot{\rho}(\mathbf{r}, t) + \rho_0 \psi(\mathbf{r}, t) = 0,$$

$$\dot{\psi}(\mathbf{r}, t) + \frac{c_0^2}{\gamma\rho_0}\nabla^2\delta\rho(\mathbf{r}, t) + \frac{c_0^2}{\gamma}\alpha\nabla^2\delta T(\mathbf{r}, t) - D_v\nabla^2\psi(\mathbf{r}, t) = K(\mathbf{r}, t),$$

$$\delta\dot{T}(\mathbf{r}, t) + \frac{(\gamma - 1)}{\alpha}\psi(\mathbf{r}, t) - \gamma D_T\nabla^2\delta T(\mathbf{r}, t) = H(\mathbf{r}, t),$$

$$(3.30)$$

where ρ_0 is the equilibrium density value, c_0 the adiabatic sound velocity ($c_0^2 = (\partial P/\partial\rho)_S$, with S the entropy), γ the heat specific ratio, $\alpha = -\rho^{-1}(\partial\rho/\partial T)_P$ the thermal expansion coefficient, D_v the longitudinal viscosity ($D_v = (\frac{4}{3}\eta_s + \eta_b)/\rho_0$ with η_s and η_b shear and bulk viscosity respectively), and D_T the thermal diffusivity ($D_T = \lambda/\rho_0 c_p$ with λ the thermal conductivity and c_p the specific heat at constant pressure). $H(\mathbf{r}, t)$ is proportional to the deposited heat due to the pump field absorbtion and $K(\mathbf{r}, t)$ represents the electrostrictive stress induced by the same pumps. These forcing terms can be expressed as functions of the total excitation electric field, which is the sum of the two pumps electric fields, through the excitation term defined in (3.10)

$$K(\mathbf{r}, t) = \frac{(\partial\epsilon/\partial\rho)_T}{8\pi}\nabla^2\left\{\text{Tr}\left[F_{kl}(\mathbf{r}, t)\right]\right\}, \qquad (3.31)$$

$$H(\mathbf{r}, t) = \frac{nc\alpha_a}{4\pi\rho_0 c_v}\text{Tr}\left[F_{kl}(\mathbf{r}, t)\right], \qquad (3.32)$$

with c the speed of light in vacuum, α_a the light absorption coefficient at the pump wavelength, and c_v the specific heat at constant volume.[6] As reported in Sect. 3.2.1.2 the TG signal is directly related to a q-component of spatial Fourier transform of the dielectric constant change, thus it is necessary to write the LHE in the q space

$$\delta\dot{\rho}(\mathbf{q}, t) + \rho_0\psi(\mathbf{q}, t) = 0,$$

$$\dot{\psi}(\mathbf{q}, t) - q^2\frac{c_0^2}{\gamma\rho_0}\delta\rho(\mathbf{q}, t) - q^2\frac{c_0^2}{\gamma}\alpha\delta T(\mathbf{q}, t) + q^2 D_v\psi(\mathbf{q}, t) = K(\mathbf{q}, t),$$

$$\delta\dot{T}(\mathbf{q}, t) + \frac{(\gamma - 1)}{\alpha}\psi(\mathbf{q}, t) + q^2\gamma D_T\delta T(\mathbf{q}, t) = H(\mathbf{q}, t).$$

$$(3.33)$$

With the condition of the ideal excitation state described in Sect. 3.2.1.2 and considering only the grating contribution, the forcing terms appearing in the above equation become

$$K(\mathbf{q}, t) = -q^2 K_0\delta(\mathbf{q} - \mathbf{q}_0)\delta(t), \qquad (3.34)$$

$$H(\mathbf{q}, t) = H_0\delta(\mathbf{q} - \mathbf{q}_0)\delta(t), \qquad (3.35)$$

with

$$K_0 = \pi^2 \left(\frac{\partial \epsilon}{\partial \rho}\right)_T \mathcal{E}_{\text{ex}}^2 \text{Tr}\,(U_{kl}) \tag{3.36}$$

$$H_0 = \frac{2\pi^2 n c \alpha_a}{\rho_0 c_v} \mathcal{E}_{\text{ex}}^2 \text{Tr}\,(U_{kl})\,, \tag{3.37}$$

where U_{kl} is the tensor defined in (3.12). Once again we want to stress that the excitation sources fix the investigated q-vector through the $\delta(\mathbf{q} - \mathbf{q}_0)$ function, so in the following equations we will understand the q value as that imposed by the experiment. By using a Laplace transform approach, (3.33) can be easily solved [37, 38] obtaining explicit expressions of $\delta\rho(q, t)$, $\delta T(q, t)$, and $\psi(q, t)$ as a function of their initial values, which, considering the $\delta(t)$ excitation of the pumps, are $\delta\rho(q, 0) = 0$, $\delta T(q, 0) = H_0$, and $\psi(q, 0) = -q^2 K_0$. Thus we get for the time evolution of the density and temperature ($t \geq 0$)

$$\delta\rho(q, t) = -q^2 G^{\rho\psi}(q, t) K_0 + G^{\rho T}(q, t) H_0, \tag{3.38}$$

$$\delta T(q, t) = -q^2 G^{T\psi}(q, t) K_0 + G^{TT}(q, t) H_0, \tag{3.39}$$

where the G^{ab} are the Green's functions of the system of (3.33). We have to note that in the TG case, the above results for the time evolution are rigorously valid for a $\delta(t)$ excitation. Otherwise, they should be convoluted with the actual shape of the exciting forces. Remembering (3.27), the dielectric constant change and thus the TG signal is

$$\delta\epsilon_{ij}(q, t) = \delta_{ij} \left(\frac{\partial \epsilon}{\partial \rho}\right)_T \left[-q^2 G^{\rho\psi}(q, t) K_0 + G^{\rho T}(q, t) H_0\right]$$
$$+ \delta_{ij} \left(\frac{\partial \epsilon}{\partial T}\right)_\rho \left[-q^2 G^{T\psi}(q, t) K_0 + G^{TT}(q, t) H_0\right]. \tag{3.40}$$

The expression of the Laplace transforms of the Green's functions are [37]

$$G^{\rho\psi} = -\rho_0 \frac{s + \gamma D_T q^2}{M(s)}$$

$$G^{T\psi} = -\frac{\gamma - 1}{\alpha} \frac{s}{M(s)}$$

$$G^{\rho T} = -\frac{\alpha \rho_0 \omega_0^2}{\gamma M(s)}$$

$$G^{TT} = \frac{s(s + D_v q^2) + \omega_0^2/\gamma}{M(s)}, \tag{3.41}$$

where $\omega_0 = c_0 q$, and $M(s) = s^3 + (D_v + \gamma D_T) q^2 s^2 + (\omega_0^2 + \gamma D_T D_v q^4) s + \omega_0^2 D_T q^2$ is the determinant of the linear system representing the hydrodynamic equations in the Laplace space. Under particular assumptions[7] the inverse Laplace transforms of responses (3.41) have an analytical expression and become, keeping only the lower order terms in q,

$$G^{\rho\psi} \simeq -\frac{1}{q}\frac{\rho_0}{c_0} e^{-q^2 \Gamma t} \sin(\omega_0 t)$$

$$G^{T\psi} \simeq -\frac{1}{q}\frac{\gamma - 1}{c_0 \alpha} e^{-q^2 \Gamma t} \sin(\omega_0 t)$$

$$G^{\rho T} \simeq -\frac{\alpha \rho_0}{\gamma} \left[e^{-q^2 D_T t} - e^{-q^2 \Gamma t} \cos(\omega_0 t) \right]$$

$$G^{TT} \simeq \frac{1}{\gamma} \left[e^{-q^2 D_T t} + (\gamma - 1) e^{-q^2 \Gamma t} \cos(\omega_0 t) \right], \qquad (3.42)$$

with[8] $\Gamma = \frac{1}{2}\left[(\gamma - 1) D_T + D_v \right] \approx \frac{1}{2} D_v$. From the expressions of the Green's functions (3.42), we see that the induced temperature and density changes relax towards equilibrium through exponentially damped acoustic oscillations and thermal exponential decays. The $\sin(\omega_0 t)$ terms describe the response to the electrostrictive pressure grating, the ones in $\cos(\omega_0 t)$ the response to the pressure grating induced by the heating, while the exponential decays are related to the vanishing of the thermal grating by heat diffusion. The Green's functions (3.42) constitute the elementary response functions (ERF) introduced in the previous section. Each G^{ab} describes the time evolution of the variable a because of the forcing term driving the variable b directly in the hydrodynamic equations (for e.g., $G^{\rho\psi}$ describes the time evolution of the variable ρ owing to the electrostrictive effect that is the forcing term of the variable ψ). Remembering expressions (3.20), (3.25), and (3.40), we can express the response function as a sum of the ERF

$$\mathcal{R}_{ijkl}^{(\mathrm{ISBS})} = -\delta_{ij}\delta_{kl} \left[\left(\frac{\partial \epsilon}{\partial \rho} \right)_T G^{\rho\psi} + \left(\frac{\partial \epsilon}{\partial T} \right)_\rho G^{T\psi} \right] q^2 A_K,$$

$$\mathcal{R}_{ijkl}^{(\mathrm{ISTS})} = \delta_{ij}\delta_{kl} \left[\left(\frac{\partial \epsilon}{\partial \rho} \right)_T G^{\rho T} + \left(\frac{\partial \epsilon}{\partial T} \right)_\rho G^{TT} \right] A_H, \qquad (3.43)$$

where A_K and A_H are amplitude coefficients function of the parameters contained, respectively, into the expressions of the two forcing terms K_0 and H_0. Not considering the birefringence effects in this hydrodynamic model, only the $\mathcal{R}_{iikk}$ components are nonzero, and are moreover all equal to each other. This simple viscoelastic model has the merit of yielding a simple analytical response from which the interesting quantities can be directly extract as free fitting parameters: sound velocity, acoustic attenuation time, and thermal diffusion time. Nevertheless, it is able to correctly describe the TG relaxational scenario of a liquid only when the structural and birefringence effects can be neglected. This occurs for all liquids, generally, at high temperatures, where the structural time is much lower than the acoustic period, and at low temperatures, where the structural time is much longer than the thermal diffusion time.

Viscoelastic model.   In the following section, we consider the generalization of LHE to the case of a molecular liquid under the hypothesis of zero birefringence effects but having the structural relaxation. This relaxation shows a complex nonexponential behavior, with relaxation times that span over many decades with decreasing of the temperature, as the sample goes from the liquid phase to the glass one. The simplest approach for taking into account these effects is to consider a time dependent longitudinal viscosity (also called memory function), as proposed by Zwanzig and Montain in 1966 to explain the light scattering spectra of molecular liquids [39,40]. Thus the product $q^2 D_{\mathrm{v}} \psi$ in the second equation of (3.33) has to be substituted with a time convolution product:

$$
\dot{\psi}(q,t) - q^2 \frac{c_0^2}{\gamma \rho_0} \delta \rho(q,t) - q^2 \frac{c_0^2}{\gamma} \alpha \delta T(q,t)
$$
$$
+ q^2 \int_0^\infty \mathrm{d}t' D_{\mathrm{v}}(t-t') \psi(t') = K(q,t). \quad (3.44)
$$

The formal solution of the new equation system is the same of (3.38) and (3.39), but with the Green's functions dependent also on the particular time dependence of the memory function. Only for particular time-dependence of $D_{\mathrm{v}}(t)$, we can obtain an analytical expression for $G^{ab}(q,t)$: for instance, when D_{v} is the longitudinal viscosity introduced by the Debye model for a viscous fluid [37–40]. In this approach $D_{\mathrm{v}}(t)$ is approximated by the sum of two terms: $D_{\mathrm{v}}(t) = D_{\mathrm{v}} \delta(t) + (c_\infty^2 - c_0^2)\tau_{\mathrm{R}} \exp(-t/\tau_{\mathrm{R}})$. This expression takes into account the memory effects in the liquid viscosity: the first term is the usual viscosity, while the second is the retarded contribution. τ_{R} is a temperature-dependent relaxation time and c_∞ and c_0 the infinite- and zero-frequency sound velocities, respectively.

If we use this expression for $D_{\mathrm{v}}(t)$ and we suppose all the characteristic relaxation times to be well separated from each other (i.e., the period of the acoustic wave, its damping time-constant, the structural relaxation time and the thermal diffusion time-constant, $q^2 D_{\mathrm{v}} \ll c_0 q$, $q^2 D_{\mathrm{T}} \ll c_0 q$, and $q^2 D_{\mathrm{T}} \gg 1/\tau_{\mathrm{R}}$), the density Green's functions, at lower order in q, are [12,36] (for $t \geq 0$)

$$
G^{\rho T}(q,t) \simeq A \left[\mathrm{e}^{-q^2 D_{\mathrm{T}} t} - \mathrm{e}^{-q^2 \Gamma_{\mathrm{A}} t} \cos(\omega_{\mathrm{A}} t) \right] + B \left[\mathrm{e}^{-q^2 D_{\mathrm{T}} t} - \mathrm{e}^{-t/\tau_{\mathrm{R}}'} \right],
$$
$$
(3.45)
$$

$$
G^{\rho \psi}(q,t) \simeq C \left[\mathrm{e}^{-q^2 \Gamma_{\mathrm{A}} t} \sin(\omega_{\mathrm{A}} t) \right], \qquad\qquad\qquad (3.46)
$$

where τ_R' is the effective relaxation time ($\tau_R' = \tau_R(c_A/c_0)^2$ with c_A the sound velocity), ω_A and $q^2\Gamma_A = 1/\tau_A$ are the frequency and damping of the acoustic longitudinal phonon; the amplitudes A, B, and C now become more complicated functions of both τ_R and c_A. In comparison with the solutions for a monatomic fluid, one more contribution exists, the relaxation mode $-Be^{-t/\tau_R'}$, moreover the parameters of the acoustic and thermal mode now depend on τ_R. These response functions allow the extraction of much information, catching the main features of the temperature and q behavior of the various parameters. In particular, they predict a maximum on the acoustic damping both versus q and T, and a gradual shift of the sound velocity from the low frequency limit c_0 to the high frequency limit c_∞, upon lowering the temperature or increasing q, as shown by the following expressions [36]:

$$\omega_A = qc_A = qc_0\sqrt{D + \sqrt{D^2 + (c_0q\tau_R)^{-2}}} \qquad (3.47)$$

$$\Gamma_A = \frac{1}{2}\left\{\left[D_v + D_T\left(\gamma - c_0^2/c_A^2\right)\right] + \frac{c_\infty^2 - c_0^2}{1 + \omega_A^2\tau_R^2}\tau_R\right\}, \qquad (3.48)$$

with $D = [c_\infty^2/c_0^2 - (c_0q\tau_R)^{-2}]/2$.

The Debye model with the single exponential memory function has some severe limitations. For highly viscous fluids, such as glass-formers in the super-cooled phase, the structural relaxation time is strongly temperature-dependent and can not be reproduced generally by a single relaxation time but needs a complete distribution function. As a consequence, more complex expressions for the longitudinal viscosity must be used, such as a multiexponential distribution of relaxation times or a stretched exponential approach [41–46]. When all the relaxation times of the actual distribution are well separated from the other characteristic times, the longitudinal viscosity affects only the second part of (3.45) and a complex relaxation can be put directly in the response as a stretched exponential, i.e. a Kohlrausch-Williams-Watts (KWW) function [36]

$$G_{\rho T}(q, t) \simeq A\left[e^{-q^2 D_T t} - e^{-q^2\Gamma_A t}\cos(\omega_A t)\right]$$
$$+ B\left[e^{-q^2 D_T t} - e^{-(t/\tau_S)^{\beta_S}}\right], \qquad (3.49)$$

where the relaxation time τ_S and the stretching factor β_S are fitting parameters for the relaxation time distribution. Clearly, in this approach ω_A and Γ_A can no longer be expressed by (3.47) and (3.48). This approximation has the advantage of yielding an easy and readable form for the response function but it does not account for possible interactions among the different relaxing mechanisms, in particular when the relaxation and thermal times become comparable. The limits of these solutions will be better explained through the analysis of *ortho-terphenyl* data reported in Sect. 3.4.2.

Relaxing viscosity as a superposition of exponential decays   In this section, we treat a generalization of the Debye relaxing viscosity of the previous paragraph. The purpose is to develop a more complex hydrodynamic model, which is able to describe the relaxational scenario of a supercooled liquid, even in the temperature range where the previous models fail. To reach this aim, a more complex function, like the stretched exponential, should be taken into account directly in the longitudinal viscosity $D_v(t)$ (3.44). The introduction of such a complex temporal function in the longitudinal viscosity prevents an analytical solution of the hydrodynamic equations and leads necessarily to a numerical procedure. When a fitting procedure of TG data is implemented on the numerical calculation, typically, it turns out to be very slow and not reliable with the consequent increasing of the fitting times. A trick exists by which it is possible to obtain a homogeneous set of linear differential hydrodynamic equations similar to (3.33) and thus have again an analytical expression of the signal. As we will see in what follows, the stretched exponential can be approximated by a superposition of a set of decaying exponential functions [18].

We start by expressing the longitudinal viscosity, $D_v(t)$, as sum of exponential decays:

$$D_v(t) = D_v^\infty \delta(t) + \Delta D_v \sum_{i=1}^{n} \frac{G_v^{(i)}}{\tau_v^{(i)}} \exp\left(-t/\tau_v^{(i)}\right) \tag{3.50}$$

or in Laplace space

$$D_v(s) = D_v^\infty + \Delta D_v \sum_{i=1}^{n} \frac{G_v^{(i)}}{1 + s\,\tau_v^{(i)}}. \tag{3.51}$$

In these equations, the n amplitudes $G_v^{(i)}$ are assumed to be normalized ($\sum_{i=1}^{n} G_v^{(i)} = 1$), leading to a normalization constant ΔD_v equals to the difference between the zero and the infinite frequency limits of the Fourier transform $D_v(\omega) = \int_0^{+\infty} dt\, e^{i\omega t} D_v(t)$: $\Delta D_v = D_v^0 - D_v^\infty$, with $D_v^0 = D_v(\omega = 0)$. We note also that with the functional form adopted for $D_v(t)$ the δ–function limit is characteristic both of the very fast and of the very slow responding medium, and that the frequency indices can be also directly interpreted in the time domain as structural time limits: in fact, considering that in the limit $\tau_v^{(i)} \to 0$ we have $\exp\left(-t/\tau_v^{(i)}\right)/\tau_v^{(i)} \to \delta(t)$, while the term is zero in the limit $\tau_v^{(i)} \to \infty$, we have

$$D_v^0 \delta(t) = \lim_{\tau_v^{(i)} \to 0} D_v(t) = \left(D_v^\infty + \Delta D_v \sum_i G_v^{(i)}\right) \delta(t) \tag{3.52}$$

and

$$D_v^\infty \delta(t) = \lim_{\tau_v^{(i)} \to \infty} D_v(t). \tag{3.53}$$

To reduce the dynamics to an equivalent set of constant coefficient equations, we define the set of auxiliary amplitudes

$$\zeta_{\mathrm{v}}^{(i)}(t) = \frac{1}{\tau_{\mathrm{v}}^{(i)}} \exp(-\frac{t}{\tau_{\mathrm{v}}^{(i)}}) \otimes \psi_q(t), \qquad (3.54)$$

with initial conditions $\zeta_{\mathrm{v}}^{(i)} = 0$ at $t = 0$. By a direct differentiation one obtains the evolution equations

$$\frac{\partial}{\partial t} \zeta_{\mathrm{v}}^{(i)}(t) = -\frac{\zeta_{\mathrm{v}}^{(i)}(t)}{\tau_{\mathrm{v}}^{(i)}} + \frac{\psi_q(t)}{\tau_{\mathrm{v}}^{(i)}}. \qquad (3.55)$$

The full set of equations can thus be written in the following compact form

$$\frac{\partial}{\partial t} \begin{pmatrix} \delta\rho_q \\ \psi_q \\ \delta T_q \\ \zeta_{\mathrm{v}}^{(1)} \\ \dots \\ \zeta_{\mathrm{v}}^{(n)} \end{pmatrix} = -\mathbf{M} \cdot \begin{pmatrix} \delta\rho_q \\ \psi_q \\ \delta T_q \\ \zeta_{\mathrm{v}}^{(1)} \\ \dots \\ \zeta_{\mathrm{v}}^{(n)} \end{pmatrix} \qquad (3.56)$$

where

$$\mathbf{M} = \begin{pmatrix} 0 & \rho_0 & 0 & 0 & \dots & 0 \\ -\frac{\omega_0^2}{\gamma\rho_0} & q^2 D_{\mathrm{v}}^\infty & -\frac{\alpha\omega_0^2}{\gamma} & q^2\Delta D_{\mathrm{v}} G_{\mathrm{v}}^{(1)} & \dots & q^2\Delta D_{\mathrm{v}} G_{\mathrm{v}}^{(n)} \\ 0 & \frac{\gamma-1}{\alpha} & \gamma D_t q^2 & 0 & \dots & 0 \\ 0 & -1/\tau_{\mathrm{v}}^{(1)} & 0 & 1/\tau_v^{(1)} & \vdots & \vdots \\ \vdots & \vdots & \vdots & \dots & \ddots & 0 \\ 0 & -1/\tau_{\mathrm{v}}^{(n)} & 0 & \dots & \dots & 1/\tau_{\mathrm{v}}^{(n)} \end{pmatrix}.$$

$$(3.57)$$

The approach is particularly suited for the fitting of a HD-TG signal, directly in the time domain, as compared with a fast Fourier approach for a longitudinal viscosity described by, e.g., a Cole-Davidson function. Indeed we can solve (3.56) by diagonalizing, with standard routines, the nonsymmetric $\mathbf{M}$ matrix, i.e. $\mathbf{M} \cdot \mathbf{V} = \mathbf{V} \cdot \mathbf{D}$ where $\mathbf{D}$ is diagonal and the resulting time solution is written as

$$\mathbf{X}(t) = \mathbf{V} \cdot \exp(-\mathbf{D}\,t) \cdot \left[\mathbf{V}^{-1} \cdot \mathbf{X}(0)\right]. \qquad (3.58)$$

Concerning the speed of the algorithm, important feature in a fitting routine, we note that it is possible to avoid the $\mathbf{V}$ matrix inversion, since $\mathbf{Y} = \mathbf{V}^{-1} \cdot \mathbf{X}(0)$ is the solution of linear system $\mathbf{V} \cdot \mathbf{Y} = \mathbf{X}(0)$. Moreover, we want to stress that expression (3.58) is analytical and only the elements of the amplitude matrix, $\mathbf{V}$, and the root matrix, $\mathbf{D}$, need to be numerically calculated.

As already stated at the beginning of the paragraph, in the glass-forming liquids, the experimental observations lead us to take the stretched exponential as representation of time dependence of the viscosity, so a proper choice of the amplitudes, $G_v^{(i)}$, and of the times, $\tau_v^{(i)}$ of the distribution have to be considered (the description of the mathematical aspects necessary for the multiexponential representation of the stretched exponential are reported in the appendix). In Sect. 3.4.3 we will report the analysis of the structural relaxation dynamics of Glycerol using such a distribution with 40 exponential decays.

3.3 Experimental Procedure

In this section, we will relate the experimental aspects of the HD-TG technique. In particular, we shall describe the laser sources producing the pump and probe beams, the optical set-up used to realize the experiment, and we will thoroughly analyze the heterodyne detected signal.

3.3.1 Laser System and Optical Setup

The lasers and the optical set-up present in our laboratory are reported in Fig. 3.3. Infrared pump pulses at $1064\,\text{nm}$ wavelength with temporal length of $20\,\text{ps}$ and repetition of $10\,\text{Hz}$ are produced by an amplified regenerated oscillator

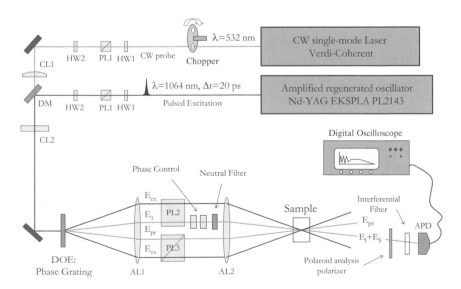

Fig. 3.3 Optical set-up and laser system for the HD-TG experiment with optical heterodyne detection: HW#, halfwave plate; PL#, polarizer; CL#, cylindrical lens; DM, dichroic mirror; DOE, diffractive optical element; AL#, achromatic lens; APD, avalanche photodiode

(Nd-YAG EKSPLA PL2143). The output pulses can reach an energy of 50 mJ. The probing beam, at 532 nm wavelength, is produced by a diode-pumped intracavity-doubled Nd-YVO (Verdi-Coherent); this is a CW single-mode laser characterized by an excellent intensity stability with low and flat noise-intensity spectrum.

The two pairs of half-wave plates and polarizers, HW1 and PL1, allow tuning of the pump and probe beam intensities and the other two half-wave plates, HW2, enable control of their linear polarization direction. Then the two beams are collinearly coupled by the dichroic mirror DR and are sent on the grating phase (DOE, diffractive optical element), described in detail later. This produces the two excitation pulses, E_{ex}, the probing, E_{pr}, and the reference beam, E_l. These beams are collected first by an achromatic lens (AL1, $f = 160$ mm), cleaned by a spatial mask to block other diffracted orders, and then recombined and focused on the sample by a second lens (AL2, $f = 160$ mm). The local laser field is also attenuated by a neutral density filter and adjusted in phase by passing through a pair of quartz prisms. The excitation grating produced on the sample is the mirror image of the enlightened DOE phase pattern. If AL1 and AL2 have the same focal length, the excitation grating has half the spacing of DOE [13, 14]. To make the polarization configuration VHVH, it is necessary to insert between the two achromatic lenses the two polarizers PL2 and PL3 and choose, with HW2, a $45°$ polarization for both pump and probe beams.

The HD-TG signal, whose polarization has been selected by a Polaroid polarizer, is optically filtered and measured by a fast avalanche silicon photodiode with a bandwidth of 1 GHz (APD, Hamamatsu). The signal is then amplified by a DC-800 MHz AVTECH amplifier and recorded by a digital oscilloscope with a 7 GHz bandwidth and a 20 Gs/s sampling rate (Tektronix). The measured signals for every sample span over many decades in time, typically up to about 1 ms, and so we record the data in a pseudo-logarithmic time scale. We use a fast time window ($0-1$ µs range with a 50 ps time-step), an intermediate ($0-20$ µs range with a 4 ns time-step) and a long one ($0-1$ ms range with a 200 ns time-step). Then the measurements, after a fine adjustment of both background and amplitude, are merged in a single data file. The large overlapping time range of the three scans permits a fine checking of their shape. No problems of overlapping occurred. Each scan is an average of 5,000 recordings, which is sufficient to produce an excellent signal to noise ratio. The temporal resolution of the experiment does not depend on the pump pulse width but only on the bandwidth of the detection system.

We reduced the laser energy on the samples to the lowest possible level to avoid undesirable thermal effects, and the CW beam has been gated in a window of about 1 ms every 10 ms by using a mechanical chopper synchronized with the excitation pulses. Depending on the sample, the mean exciting energy was in the range $0.4-4$ mJ for each pump pulse and the probing power was about

10−100 mW. The reference beam intensity is very low; it was experimentally set by means of a variable neutral filter in order to be almost 100 times the intensity of the diffracted signal. With these intensities, the experiment was well inside the linear response regime and no dependence of HD-TG signal shape on the intensities of the beams could be detected.

As already stated in the introduction, the diffractive optical element DOE was first introduced in 1998 [13,14]. It provides considerable advantages. Essentially a grating phase, characterized by a square shaped profile, is hollowed out on a fused silica plate by ion beam techniques. Thanks to the square profile of the grating, it is possible to obtain very high diffraction efficiency on the only first orders by controlling the depth Δ of the grooves. Choosing $\Delta = \frac{\lambda/2}{(n-1)}$, with n the refractive index of silica, the diffraction efficiency would be, theoretically, equal to 50% on each first order. Since we have to diffract both 1064 and 532 nm beams, a compromise must be reached. The chosen DOE (made by Edinburgh Microoptics) is optimized for 830 nm radiation, and it gives on a single beam at first order a 12% diffraction efficiency for the 532 nm and 38% for the 1,064 nm.

Some important features have to be emphasized. This type of set-up automatically gives the Bragg condition on all the beams and produces a very stable phase locking between the probing and reference beam, a crucial parameter to realize HD. Moreover, by choosing different spacing of grooves, it is possible to easily change the angle between the exciting beams and consequently the q-vector induced in the sample. For the latter reason, we use a DOE composed of six phase gratings characterized by different spacings. They produce on the sample the following q-vector values: 0.630, 1.00, 1.38, 1.79, 2.10, and 2.51 μm^{-1} (if AL1 and AL2 have the same focal length). Other values of q can be obtained changing the focal length of AL2 (e.g. 0.338 μm^{-1} has been obtained with $f_{AL1} = 160$ mm and $f_{AL2} = 300$ mm).

To properly measure the natural damping of the induced acoustic oscillations [11, 12], the pump and probe beams are focused on the DOE following the geometry sketched in Fig. 3.4. The pump is focused by the cylindrical lens, CL2, in order to have on the sample an excitation grating extended in the q-direction (about 0.5×5 mm) while the probe beam is focused through the two crossed cylindrical lenses CL1 and CL2 to a circular spot in order to have a probing area with much smaller dimensions in the q-direction (0.5×0.5 mm). Without this excitation-probing geometry, an acoustic damping due to geometrical reasons could be probed. In fact, the probed stationary wave obtained as superposition of the two induced counter-propagating acoustic waves may vanish, owing to the escaping from the probing area of these two waves. Figure 3.5 shows a comparison between two HD-TG data on methanol at 300 K and at $q = 0.630$ μm^{-1} performed with two different extensions in the q-direction of the pump spot size and with the same circular probe spot size of 0.5 mm.

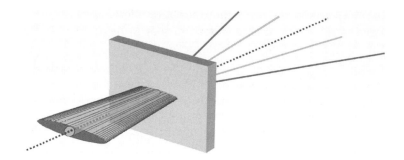

Fig. 3.4 The cylindrical spot shape of pump and the circular spot of probe on the DOE. The same spot beam image is reproduced on the sample by the couple of lenses AL1 and AL2

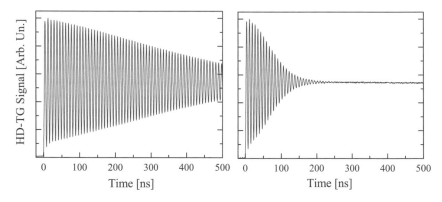

Fig. 3.5 Comparison between two HD-TG data on methanol at 300 K and at $q = 0.630 \, \mu m^{-1}$ performed with two different extensions in q-direction of the pump spot size: $\sim$5 mm (left graph) and $\sim$1 mm (right graph), and with the same circular probe spot size of 0.5 mm. The damping observed in the acoustic oscillations is clearly different in the two cases

In the left graph the pump extension is about 5 mm, while in the right graph it is reduced to about 1 mm. Clearly the envelope of acoustic oscillations is different: in the first case, we probe almost exclusively the natural damping, whereas in the second the damping is dominated by the gaussian spatial shape of the pumps.

3.3.2   Heterodyne and Homodyne Detections

Typically, detectors of electromagnetic field measure the intensity:

$$S\left(t\right) = I_{\mathrm{s}}\left(t\right) = \left\langle |\mathbf{E}(t)|^{2} \right\rangle_{\mathrm{op.c.}},
\tag{3.59}$$

where $\langle \cdot \rangle_{\mathrm{op.c.}}$ means the time averaging over the optical period.[9] A direct measurement of the scattering intensity is called homodyne detection (HO). If the measured field is a signal field supplying dynamic information about some relaxing system, like in a TG experiment, through a response function $\mathcal{R}(t)$ (i.e. $E_s(t) \propto \mathcal{R}(t)$), the homodyne signal is clearly proportional to the square of the response function.

Another detection mode exists through which it is possible to directly measure the amplitude of the diffracted field; this detection mode is called optical heterodyne detection. In HD, both the signal and a reference field are superimposed on the detector, and the intensity of interference field is recorded. Usually the diffracted field can be written as (see Sect. 3.2.1.2) $\mathbf{E}_s(t) = \hat{e}\mathcal{E}_s(t) \exp[i(\mathbf{k}_s \cdot \mathbf{r} - \omega t)] + c.c.$, with $\mathcal{E}_s(t)$, in general, a complex function; then we choose a reference field with same wave-vector, direction and frequency of $\mathbf{E}_s(t)$ and with constant amplitude $\mathbf{E}_l(t) = \hat{e}\mathcal{E}_l \exp[i(\mathbf{k}_s \cdot \mathbf{r} - \omega t + \varphi)] + c.c.$, where φ is the optical phase between the signal and reference field and $\mathcal{E}_l$ is a real amplitude. Hence, the heterodyne signal is

$$S(t) = \left\langle |\mathbf{E}_s(t) + \mathbf{E}_l|^2 \right\rangle_{\text{op.c.}} \tag{3.60}$$
$$= I_s(t) + I_l + 2\{\text{Re}[\mathcal{E}_s(t)]\mathcal{E}_l \cos\varphi + \text{Im}[\mathcal{E}_s(t)]\mathcal{E}_l \sin\varphi\}.$$

The first two terms in the right-hand side of (3.60) are the homodyne $I_s(t) = \left\langle |\mathbf{E}_s(t)|^2 \right\rangle_{\text{op.c.}}$ contribution and the local field intensity I_l, while the third, between curly braces, is the heterodyne contribution. If the local field has a high intensity, the homodyne contribution becomes negligible and the time variation of the signal is dominated by the heterodyne term, which is directly proportional to the signal field. Moreover, this last term can be experimentally isolated by subtracting two signals characterized by different phases. By recording a first signal, S_+, with $\varphi_+ = \varphi_0$ and then a second one, S_-, with $\varphi_- = \varphi_0 + \pi$, it immediately follows

$$S^{\text{HD}}(t) = [S_+ - S_-] \tag{3.61}$$
$$= 4\{\text{Re}[\mathcal{E}_s(t)]\mathcal{E}_l \cos\varphi_0 + \text{Im}[\mathcal{E}_s(t)]\mathcal{E}_l \sin\varphi_0\}.$$

In a TG experiment, the dynamic birefringence-phase and dichroic-amplitude gratings are linked to $\text{Re}[\mathcal{E}_s(t)]$ and $\text{Im}[\mathcal{E}_s(t)]$, respectively [1, 47, 48]. It is clear from expression (3.61) that by choosing $\varphi_0 = 0$ or $\pi/2$, it is possible to extract only the real or imaginary part of $\mathcal{E}_s(t)$ and then to measure only the dichroic-amplitude or the birefringence-phase contribution.

When the dichroic-amplitude grating is negligible, $\mathcal{E}_s(t)$ is real and the heterodyne term is $\mathcal{E}_s(t)\mathcal{E}_l \cos\varphi$, then choosing $\varphi_+ = 2n\pi$ and $\varphi_- = (2n+1)\pi$ (n integer) we get

$$S^{\text{HD}}(t) = [S_+ - S_-] = 4\mathcal{E}_s(t)\mathcal{E}_l. \tag{3.62}$$

It is important to emphasize the major advantages in using heterodyne instead of HO. First, it substantially improves the signal to noise ratio in the observed

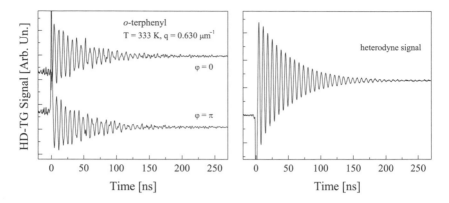

Fig. 3.6 Typical HD-TG raw data corresponding to a phase difference between signal and local reference of $\varphi = 0$ and $\varphi = \pi$ (left graph); pure HD-TG signal (right graph) is obtained by subtracting the two previous signals. The subtraction enables the improvement of the signal/noise ratio, removing the homodyne and all the phase-independent spurious contributions

time window, because of both the signal increment and the cancellation of the spurious signals that are not phase sensitive. Second, it enhances the dynamic range since the recorded signal is directly proportional to signal field instead of being proportional to its square. In the study of materials with a weak scattering efficiency and complex responses, these features turn out to be of basic importance. Furthermore, HD allows the measurement of the signal at very long times, where the TG signals become very weak. Nevertheless, the effective realization of such detection is quite difficult at optical frequencies. Indeed, to get an interferometric phase stability between the diffracted and the local field is not a simple experimental task, which explains why only very few HD-TG experiments have been realized so far [1]. The introduction of the phase mask DOE in the TG optical setup (see Sect. 3.3.1) has considerably reduced the difficulties of achieving HD. We report in Fig. 3.6 a typical HD-TG result. The pure heterodyne signal, S^{HD}, is extracted from the two S_+ and S_- signals corresponding to the phases $\varphi_+ = 0$ and $\varphi_- = \pi$, respectively. The figure shows how the HD substantially increases the quality of data.

3.3.3   Fitting Procedure

To compare our data to the theoretical responses, we need to take into account the response of detection system by convoluting the theoretical signal with the instrumental function. This function has been obtained by sampling the response of the detection system to the second harmonic of the pump infrared pulse. We adopted this procedure because we needed to know the real response of the detection system at the probing wavelength of 532 nm. This response is different from that at 1064 nm because silicon photodiodes show a response in the visible spectrum faster than in the infrared region. This procedure is correct as long as the width of the signal pulse depends only on the bandwidth of detection system (photodiode, amplifier, and oscilloscope), and not on the temporal length or shape of exciting pulse.

Many samples present also an instantaneous electronic response at zero time due to the hyperpolarizability response. This contribution has a shape exactly equal to that of the instrumental function and, by covering only the first ns of the signal, it does not affect the analysis of the interesting dynamics. However, we include this contribution anyway by adding to the theoretical signal S_{th} (which describes only the nuclear response) a δ-time function and then making the convolution with the instrumental function $I(t)$. The fitting function is thus

$$S = \int_{-\infty}^{\infty} \left[A\delta\left(t - t'\right) + S_{th}\left(t - t'\right) \right] I\left(t'\right) \, dt', \qquad (3.63)$$

where A is an amplitude factor. The nonlinear least square fit of our data are performed by a numerical routine based on a *modified Gaussian-Newton* algorithm.

3.4 Experimental Results

In this section we report some HD-TG results obtained on different liquids. The majority of these data have already been published, so here we want to review under unified point of view the various interpretations. We will give at first a general description of the relaxation dynamics measured by a TG experiment and then we shall analyze some TG data obtained on different liquids. We will draw attention to the different hydrodynamic models that must be used to correctly describe some of the complex aspects of the supercooled-liquid relaxation dynamics.

As outlined in Sect. 3.2 in a TG experiment performed on molecular liquids, the refractive index change is produced by three excitation channels. First, the pump light is partly absorbed generating a local heating and consequently a density change (ISTS). Second, it creates an instantaneous electrostrictive pressure generating again a density perturbation (ISBS). Third, the excitation pulse, through its electric field, orients the molecules inducing a birefringence in the sample (PIB). Moreover, we know that by means to the R–TC effect, the density grating, induced by the heating and electrostriction, produces a molecular velocity grating and then a birefringence grating. By the same coupling, the PIB grating generates longitudinal or transverse phonons. The magnitude of these contributions depends on the level of pump absorption and on the polarizability level of the molecules.

Many molecular liquids show a weak absorption at the pump wavelength 1064 nm, due to overtones and/or combinations of the vibrational bands. Typically, these vibrational excitations thermalize in a few picoseconds, producing immediately a thermal grating. Despite the weakness of the absorption effect, the ISTS contribution is often the most important part of the TG signal. Normally, the ISBS contribution is also present in the signal. The strength of ISBS is directly defined by the intensity of the isotropic part of the molecular polarizability. So polarizable molecules, such as *ortho*-terphenyl (OTP), show a significant ISBS signal, whereas TG experiments on less polarizable liquids such as glycerol are dominated by the ISTS signal.

Typical HD-TG data are shown in Fig. 3.7. These, generally, display damped oscillations produced by acoustic phonons superimposed on a slower decay generated by the thermal processes. In particular, the figure shows three examples where the ISTS and ISBS contributions count differently. The reported HD-TG data are taken on CCl_4, glycerol, and OTP. Because of the negligible absorption at the pump wavelength, CCl_4 shows only damped acoustic oscillations due to the electrostrictive effect. Glycerol, instead, presents no electrostrictive contribution, and only the heating effect triggers the response. OTP, on the other hand, is an example where both ISBS and ISTS are present. The HD-TG data on glycerol and OTP also show a rise of the signal, due to a structural relaxation, which appears in the supercooled liquids as the temperature is decreased.

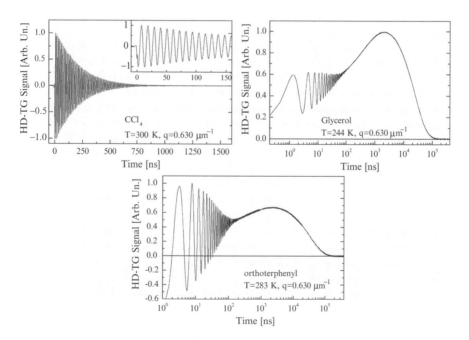

Fig. 3.7 HD-TG data on CCl_4, glycerol and OTP. The CCl_4, because of negligible absorption at the pump wavelength, shows only ISBS contribution, glycerol only ISTS one, while OTP presents both ISBS and ISTS contributions. The rise of the signal in the data on glycerol and OTP is related to the structural mode typical of supercooled liquids

The possible presence of birefringence contributions can be verified in a TG experiment by the selection of the polarizations of pump, probe, and diffracted beams. It is possible, for instance, with the polarization configuration VH of the pump pulses (see notations of Fig. 3.1), to excite a polarization grating and to take out only the PIB contribution; moreover, by changing the probe and detection polarizations from VV to HH, we can reveal the presence of a bire-fringence grating. The polarization configurations that have been used in our experiment are VVVV, HHHH, VHVH, VVHH, and HHVV, where the first two letters refer to the probe and detection polarizations respectively, and the second ones to polarizations of the two pump pulses.[10] In all of the samples (glycerol, *m*-toluidine, OTP, salol, and water), whatever their temperature, the HD-TG signal does not show any particular difference by changing the polar-izations of the pumps, which is $S_{VVVV} = S_{VVHH}$ and $S_{HHHH} = S_{HHVV}$, and the signal S_{VHVH} appears always negligible. This means that the PIB contri-bution is not observed in all the samples analyzed. By changing the probe and diffracted beam polarizations, however, the samples show different behaviors. A few samples have a negligible birefringence signal; this means that the R–TC is negligible too, while other ones show a strong dependence on the probe and diffracted beams polarizations when the structural mode starts to appear, showing a considerable R–TC effect. In Fig. 3.8, we report the HD-TG data on glycerol and OTP, as examples of media where the induced birefringence grat-ing is negligible, while in Fig. 3.9 the HD-TG data on salol and metatoluidine clearly show how the T–RC can be important.

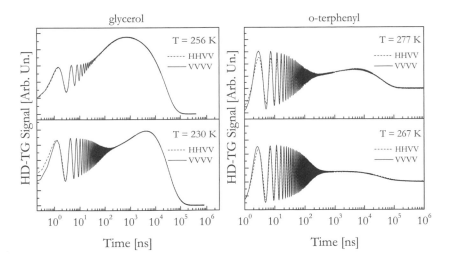

Fig. 3.8 HD-TG data from glycerol (left graph) and OTP (right graph) at $q = 0.630\,\mu m^{-1}$ for the two polarization configurations VVVV and HHVV at two different temperatures. The signal S_{VVVV} and S_{HHVV} do not show any essential difference in both the samples showing a negligible rotation–translation coupling

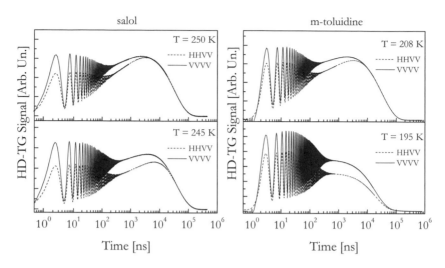

Fig. 3.9 Signals S_{VVVV} and S_{HHVV} collected on salol (left graph) and *m*-toluidine (right graph) at $q = 0.630\,\mu m^{-1}$ at two different temperatures. The S_{VVVV} and S_{HHVV} data show clearly a different profile of the signal

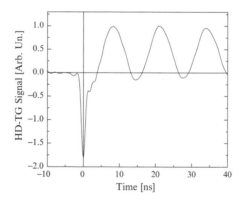

Fig. 3.10 The instantaneous response at zero time for the OTP data at $T = 333\,K$ and $q = 0.338\,\mu m^{-1}$. This contribution is connected to the molecular electronic hyperpolarizability. The latter has an intensity and a sign that depends on the polarization of the pump and probe beams and a temporal shape given by the instrumental function

In many samples, an instantaneous electronic response at zero time is present too. This is due to by the interaction of the pump fields with the electronic hyperpolarizability of the molecules; it has an intensity and a sign depending on the polarization of the pump and probe beams. Since the electronic hyperpo-larizability has an instantaneous response, this signal has exactly the same shape as the instrumental function. Figure 3.10 shows the instantaneous response at zero time for the OTP data at $T = 333\,K$ and $q = 0.338\,\mu m^{-1}$. Since this contribution covers only the first ns of the decay, it does not affect the analysis of the interesting dynamics and, therefore, we will never show it in our data even if present.

3.4.1 Supercooled Water: the Importance of the Photothermal Effect

Here, we want to report the HD-TG experiment data obtained on liquid and supercooled water [21, 22] to show when the dielectric constant change induced directly by the temperature has to be taken account. As already stated, this contribution has been generally neglected in the TG and LS signal analysis, thanks to the fact that $(\partial \epsilon / \partial T)_\rho \delta T$ is much lower than $(\partial \epsilon / \partial \rho)_T \delta \rho$ for most liquids. Water is a particular sample where this condition is not satisfied. Moreover, in the TG water signal, the temperature contribution shows itself clearly, thanks to the anomalous temperature behavior of the density, which has into a maximum around 4°C. As a consequence, the thermal expansion coefficient vanishes at the same temperature. Let us consider expression 3.27 of the dielectric constant change, at which the TG signal is proportional. If we focus our attention on the TG slow dynamics, where the pressure grating disappears, we have, from simple thermodynamic considerations, $\delta p = (\delta \rho + \alpha \rho_0 \delta T) c_0^2 / \gamma = 0$. From this relation we get $\delta \rho = -\alpha \rho_0 \delta T$ and we find that the density contribution disappears at the same temperature, highlighting the δT-contribution. This phenomenon can be understood better by observing the experimental results. In Fig. 3.11, we report some HD-TG data at $q = 0.630 \, \mu m^{-1}$ in the polarization configuration VVVV (the birefringence effects were found to be negligible at all investigated temperatures). As expected, the data show damped oscillations produced by acoustic phonons superimposed on the slow exponential decay generated by the thermal processes. As shown by the figure, the thermal signal reduces with decreasing temperature, following the temperature dependence of α, but does not vanish at 4°C since it is the sum of the two above-mentioned contributions. It becomes zero at about 0°C and changes sign for lower temperature in the supercooled phase.

Let us try to understand, now, why the thermal signal vanishes at 0°C. On the basis of simple thermodynamic considerations, we can express the density and temperature derivatives of ϵ using the refractive index, n, its pressure and temperature derivatives, and the isothermal compressibility, χ_T:

$$\left(\frac{\partial \epsilon}{\partial \rho} \right)_T = \frac{2n}{\rho_0 \chi_T} \left(\frac{\partial n}{\partial P} \right)_T, \tag{3.64}$$

$$\left(\frac{\partial \epsilon}{\partial T} \right)_\rho = 2n \left[\left(\frac{\partial n}{\partial T} \right)_P + \frac{\alpha}{\chi_T} \left(\frac{\partial n}{\partial P} \right)_T \right]. \tag{3.65}$$

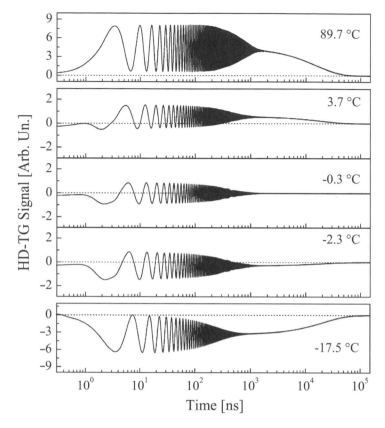

Fig. 3.11 HD-TG data on water at $q = 0.630 \ \mu m^{-1}$. The data, as usually, show damped acoustic oscillations with and a slow thermal exponential decay. The amplitude of the thermal mode shows a peculiar temperature behavior: it vanishes at temperature of about $0°C$ and changes sign for lower temperatures in the supercooled phase

If we substitute (3.64) and (3.65) in the expression (3.27) of $\delta\epsilon$, and we use the relation $\delta\rho = -\alpha\rho_0\delta T$ valid for long times, the pressure derivative terms cancel each other and we have

$$\delta\epsilon = 2n \left(\frac{\partial n}{\partial T}\right)_P \delta T. \tag{3.66}$$

For water, the refractive index as a function of temperature has a maximum around $0°C$ and hence the multiplicative factor in (3.66) reduces the signal to zero around this temperature. This simple expression of the TG signal explains the disappearance of the slow dynamics in the TG experimental results.

Let us see, now, when the $(\partial\epsilon/\partial\rho)_T\delta\rho$ term can be neglected in a TG experiment. Consider the simplest hydrodynamic model of Sect. 3.2.2.1. Since, in the temperature range analyzed, the structural relaxation time of water (some ps [49, 50]) is always faster than the other material responses ($\approx 1/\omega_0 \geq 1$ ns, being $\omega_0 = c_0 q$) and the birefringence effects are negligible, this model is suitable to describe our relaxational dynamics. Remembering (3.40) and considering, from the expressions of the Green's functions (3.42), that $G^{TT} \sim G^{\rho T}/\alpha T_0$ and $G^{T\psi} \sim G^{\rho\psi}(\gamma - 1)/\alpha T_0$, we get

$$\delta\epsilon(t) = \left[\left(\frac{\partial\epsilon}{\partial\rho}\right)_T G^{\rho T}(t)\right]\left\{1 + O\left(\frac{E}{\alpha T_0}\right)\right\} H_0$$
$$- \left[\left(\frac{\partial\epsilon}{\partial\rho}\right)_T G^{\rho\psi}(t)\right]\left\{1 + O\left(\frac{E(\gamma - 1)}{\alpha T_0}\right)\right\} q^2 K_0, \quad (3.67)$$

where we have introduced the dimensionless ratio between the dielectric constant derivatives

$$E = \frac{T_0 \left(\frac{\partial\epsilon}{\partial T}\right)_\rho}{\rho_0 \left(\frac{\partial\epsilon}{\partial\rho}\right)_T}.$$

Thus we see that $E \ll \alpha T_0$ is the condition for neglecting the dielectric constant change directly induced by temperature.

To extract the interesting parameters of water, like the values of the sound velocity, the acoustic damping time and the thermal diffusivity, we performed a least-square fit with the analytical function obtained substituting the Green's functions (3.42) in (3.40)

$$S^{HD} \propto A \, e^{-q^2\Gamma t} \cos(\omega_0 t) + B \, e^{-q^2\Gamma t} \sin(\omega_0 t) + C \, e^{-q^2 D_T}. \quad (3.68)$$

The amplitudes A, B, and C assume the following expressions:

$$A = \frac{\rho_0 H_0}{\gamma T_0}\left(\frac{\partial\epsilon}{\partial\rho}\right)_T [\alpha T_0 + E(\gamma - 1)]$$

$$B = \frac{q K_0 \rho_0}{c_0 T_0 \alpha}\left(\frac{\partial\epsilon}{\partial\rho}\right)_T [\alpha T_0 + E(\gamma - 1)]$$

$$C = -\frac{\rho_0 H_0}{\gamma T_0}\left(\frac{\partial\epsilon}{\partial\rho}\right)_T (\alpha T_0 - E). \quad (3.69)$$

The free fitting parameters were, clearly, the above amplitudes, the sound frequency ω_0, the damping rate Γ, and the thermal diffusivity D_T. The acoustic relaxation time of water, in this range of temperature and at this q-value, is very long, around 1 μs. For this reason, the attenuation of the acoustic oscillation in the data is not due only to the natural sound damping, but is also suffering from the finite size of excited grating (see Sect. 3.3.1). The decay of the sound waves depends also on the pump shape in the q-direction. In the approximation of gaussian pump beams, the sound attenuation becomes

$$e^{-q^2\Gamma t} \rightarrow e^{-t^2/\sigma_t^2} e^{-q^2\Gamma t}, \quad (3.70)$$

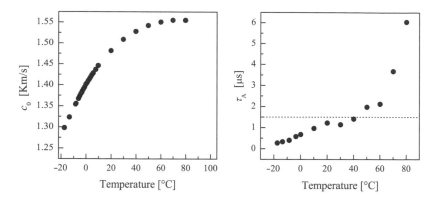

Fig. 3.12 Temperature dependence of the adiabatic sound velocity of water (left graph) and of the acoustic relaxation time (right graph). The *dottedline* represents the acoustic transit time defined by the experimental geometry, $\tau \simeq 1.5\,\mu$sec; above this limit, the reported coefficients are affected by large errors

where $\sigma_t = \sigma/c_0$ with σ the width of the pump laser spot in the q-direction. So, σ_t is the time spent by an acoustic wave to cover a distance σ. Thus, we used the fitting function (3.68) applying the substitution (3.70). σ_t was a free fitting parameter.

In Fig. 3.12, we report the temperature dependence of the adiabatic sound velocity, $c_0 = \omega_0/q$, and the acoustic damping time, $\tau_A = 1/q^2\Gamma$. As shown by the scattering in the damping data points, the fitting procedure is able to extract a valuable damping time as long as it reaches the acoustic transit time σ_t, which is represented by the dotted line at $\tau \simeq 1.5\,\mu$sec. The values of the thermal diffusivity, D_T, as a function of temperature are shown in the left graph of Fig. 3.13.

The vast literature on water makes available the values of practically all the thermodynamic parameters appearing in the hydrodynamic equations for almost the full temperature range analyzed [22]. In particular, we can calculate the E-ratio by means of (3.64) and (3.65) and estimate the $E(\gamma - 1)$ term in the amplitudes A and B, which has been shown to be always negligible. With this approximation, the E values can be simply extracted from the experimental values of the amplitudes A and C knowing only the literature data of α:

$$E \simeq \alpha T_0 \left(1 + \frac{C}{A}\right). \tag{3.71}$$

So far, we considered the simple viscoelastic hydrodynamic model with the approximations leading to relatively simple analytical expressions for the amplitudes A, B, and C. In [22], instead, we followed a different approach. We started from the same hydrodynamic model but we got the fitting function by a numerical solution of the inverse Laplace transforms of the Green's functions

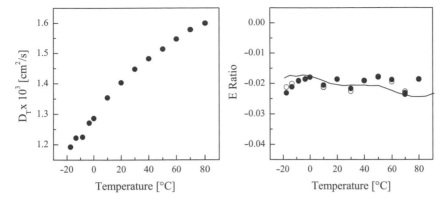

Fig. 3.13 Left graph: Temperature dependence of thermal diffusivity of water. Right graph: Temperature behavior of the E-ratio. The *full circles* refer to the experimental values calculated with (3.71) while the *open circles* to the values extracted from the full model. The experimental data are compared with ones calculated from the literature data (*solid line*) following (3.64) and (3.65)

(3.41), without any approximation. The final fitting function can be still cast in the format (3.68), but the analytical expressions of the amplitude coefficients are now quite complicated. In this case, the solution requires knowledge of many thermodynamic parameters appearing in the hydrodynamic equations but the E-ratio is directly a fitting parameter.

In Fig. 3.13, we report the E-ratio estimate obtained from the experimental values of C/A and the literature data of α following the expression (3.71) together with the values of the free fitting parameter E found with the full model. The two approaches give quite similar results confirming that the simple model approximations are valid and do not affect the evaluation of the E-ratio. Finally, both the experimental results were compared with those calculated from the literature data using formulas (3.64) and (3.65). The agreement is fairly good within the uncertainty, estimated to be $\approx 20\%$.

3.4.2 *Ortho*-terphenyl: the Structural Dynamics of a Fragile Glass-Former

We report, here, the TG experiment results obtained on *ortho*-terphenyl as a representative example of the relaxation dynamics study on a typical glass-former liquid [15]. Moreover, we chose this sample to highlight the merits and failings of the generalized hydrodynamic model of Section 3.2.2.1.

Ortho-terphenyl is a prototype of fragile glass-formers, in which it is possible to observe a wide variation of the structural relaxation time within a temperature range of only a few tens of degrees. Thanks to its melting point temperature of 329 K and a glass transition temperature of 244 K, it presents a supercooled

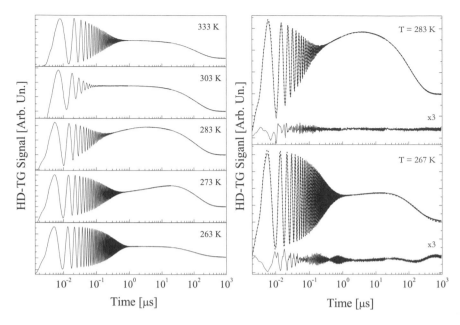

Fig. 3.14   Left graph: HD-TG data on OTP at $q = 0.338$ μm^{-1} at several temperatures from the liquid phase to the supercooled one. Right graph: HD-TG data (*solid lines*), fits (*dotted lines*) and residues ×3 (*lower lines*) at the two temperatures $T = 283$ K and $T = 267$ K at the same q-vector

phase at relatively high temperature range. However, it lacks high scattering efficiency and tends to easily crystallize.

The relaxation processes of OTP have been measured for three different q values: $q = 0.338$, 0.630, and 1.00 μm^{-1}, in the temperature range 243–373 K. Thanks to the negligible birefringence effects, we investigated only the polarization configuration VVVV. In the left graph of Fig. 3.14, we report a few HD-TG data on glass-forming OTP in a linear-log scale. From the figure, we can clearly identify three main dynamical processes in the density dynamics of OTP: a damped acoustic oscillation, a structural relaxation that appears as a rise in the signal, and finally the relaxation of the thermal grating by heat diffusion. As visible from the data, the acoustic and the structural processes are strongly temperature-dependent and they have separated time scales only in a particular range of temperatures.

To extract the interesting parameters, we perform the least-squares fit using the hydrodynamic model of Sect. 3.2.2.1. Considering $(\partial\epsilon/\partial T)_\rho\delta T \ll (\partial\epsilon/\partial\rho)_T\delta\rho$ and the birefringence effects to be negligible, we used only the density response function to extract the information about OTP dynamics from HD-TG data (see (3.40), (3.45), and (3.49))

$$S^{\mathrm{HD}}(q,t) \propto -q^2 G^{\rho\psi}(q,t)K_0 + G^{\rho T}(q,t)H_0 \qquad (3.72)$$

$$G^{\rho\psi}(q,t) \simeq C\left[e^{-q^2\Gamma_A t}\sin\left(\omega_A t\right)\right] \tag{3.73}$$

$$G^{\rho T}(q,t) \simeq A\left[e^{-q^2 D_T t} - e^{-q^2\Gamma_A t}\cos\left(\omega_A t\right)\right]$$
$$+ B\left[e^{-q^2 D_T t} - e^{-(t/\tau_S)^{\beta_S}}\right]. \tag{3.74}$$

The fitting parameters were the acoustic frequency ω_A and the acoustic relaxation time $\tau_A = 1/q^2\Gamma_A$, the structural relaxation time τ_S and the stretching parameter β_S, the thermal relaxation time $\tau_H = 1/q^2 D_T$ and finally the amplitude constants A, B, and C. Two examples of fit comparisons are reported in Fig. 3.14. From the data analysis, we find the presence of both the electrostrictive effect (ISBS) and the thermal effect (ISTS). We can estimate about 60% of ISBS against 40% of ISTS for $q = 0.338 \ \mu m^{-1}$ and this ratio increases when the value of the wave-vector increases owing to the q dependence of the ratio $q^2 G^{\rho\psi}/G^{\rho T}$. Nevertheless, the two contributions can be safely disentangled from the fits, thanks to the linear access to the response function. In Fig. 3.15, we show the sound velocity, $c_A = \omega_A/q$, and the damping rate, $1/\tau_A$. These two parameters are extracted with very small uncertainties, less than 1%, for all the investigated temperatures and wave-vectors. The sound velocity (left graph of Fig. 3.15) exhibits the typical temperature dependence of viscoelastic liquids. The sound velocity increases with decreasing temperature and passes through two different linear regimes. At high temperatures, the velocity almost corresponds to the adiabatic sound velocity, c_0, while at low temperatures, it corresponds to the solid-like or infinite frequency sound velocity, c_∞. In the transition region, we have the so-called dispersive regime where the sound velocity shows a rapid increase of its value towards c_∞. For higher wave-vectors,

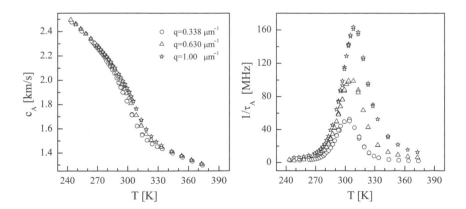

Fig. 3.15 Sound velocities c_A (left graph) and acoustic damping rates $1/\tau_A$ (right graph) versus temperature from fits of HD-TG data at the three investigated wave-vectors. The maximum of the dispersion effect occurs at the temperature where $\omega_A \tau_S \sim 1$

the dispersive regime shifts toward higher temperatures. This behavior is due to the rapid variation of the structural relaxation time with the temperature. In fact, at high temperatures, the structural relaxation times are shorter than the phonon oscillation period, $\tau_S \ll (\omega_A)^{-1}$; therefore, the acoustic phonon is weakly coupled with the structural processes and it shows a soft damping. Again at low temperatures, when $\tau_S \gg (\omega_A)^{-1}$, the two processes are decoupled yielding again a soft damping of the sound waves. Otherwise, when $\tau_S\omega_A \sim 1$, the structural and acoustic phenomena have the maximum coupling and this produces a maximum in the damping rate and in the sound velocity dispersion effect (see right graph of Fig. 3.15).

By further decreasing temperature, the structural relaxation process appears in the HD-TG data as a rise of the thermal contribution (Fig. 3.14). In this temperature range, the structural mode is well separated from the other two modes and its time can be easily extracted from the fitting procedure with confidence. Indeed, it has been possible to get reliable structural parameters, τ_S and β_S, only in a q-dependent restricted range of temperature. This fact reflects the limitation of the fitting formulas, (3.73) and (3.74), since they fully apply only when a time scale separation exists among the various characteristic times, as already pointed out in Sect. 3.2.2.1. The structural relaxation time and the β stretching parameter, as functions of temperature, are reported in Fig. 3.16. They cover three decades in time that were not previously investigated, from about 10 ns up to 20 μs, and they are indeed quite difficult to reach by other techniques, both in (q, t) and (q, ω) space. Within the uncertainties, both the structural relaxation times and the stretching parameters do not show any q-dependence. Moreover, no temperature dependence of the stretching parameters is evident.

The thermal relaxation time τ_H defines the final decay of the HD-TG signal and it is safely extracted when the condition $\tau_H \gg \tau_S$ is verified. In Fig. 3.17a),

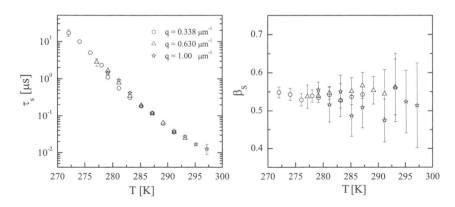

Fig. 3.16   Structural relaxation times τ_S and stretching parameter β_S versus temperature at the three wave-vectors analyzed

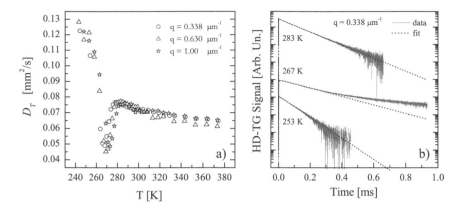

Fig. 3.17 (a) Thermal diffusivity as a function of temperature for each q-value. (b) HD-TG data (*solid curves*) and fits (*dotted curves*), at wave-vector $q = 0.338\ \mu m^{-1}$ at three different temperatures. The middle curve is at the temperature where the thermal diffusivity shows the anomalous dip and is no longer reproduced by the fitting function at long times

we report the thermal diffusivity, $D_T = (1/\tau_H q^2)$, for the three wave-vectors investigated in the whole temperature range. In the high temperature range, D_T shows the expected smooth variation and the independence of q. As the temperature is decreased, an anomalous strong decrease in the thermal diffusivity appears. The change starts at higher temperatures for higher q-values. This effect is a fitting artefact: in this temperature range, the used response function (3.73, 3.74) is not able to reproduce the data a long times. Here, in fact, the data show a strong nonexponential behavior, as clearly visible from the semilogarithmic plot of Fig. 3.17b). The slow relaxation at $T = 283$ K and 253 K are characterized by a single exponential decay, the thermal decay, while at 267 K some other relaxation appears. This effect has been ascribed to an interaction between the structural and thermal relaxation times, τ_S and τ_H, which are getting closer with decreasing temperature [16, 17]. However, when the thermal diffusivity starts deviating, the structural and thermal relaxation times seem to be very different. In fact, at $q = 0.338\ \mu m^{-1}$, and at the temperature where the deviation starts to appear, at about $T = 275$ K, we measure $\tau_S = 15\ \mu s$ and $\tau_H = 115\ \mu s$. We have to consider that τ_S is a relaxation time representative of a large distribution of times and thus the overlapping temperature range of the modes could be much wider. The subsequent large increase of the thermal diffusivity occurring at low temperature after the dip is anomalous too. In [30,31], it is suggested that this increase comes from the frequency dependence of specific heat, so that, when the structural relaxation time is much longer than the thermal diffusion time, the "effective" thermal relaxation time changes from its thermodynamic value of the high temperatures to the infinity frequency value.

We would like to stress once again that the highest limit of this simple hydrodynamic model is on the failing to properly describe the slow thermal decay at relatively low temperature. We have found this behavior common to other several glass-formers: glycerol, *m*-toluidine, salol, and all the studied glass-formers [16–18, 24, 30]. As we shall see in the next section, a more suitable hydrodynamic treatment for the structural relaxation and the heat transport should be able to take into account these complex relaxation patterns.

3.4.3   Glycerol: A Test of Hydrodynamic Models

Because of the abundance of literature data, HD-TG measurements on glycerol represent a very stringent test on the theoretical models. This substance, classified as a glass-former with an "intermediate" behavior in terms of its temperature dependence of viscosity [51], has a melting point temperature $T_m \approx 292$ K and a glass temperature $T_m \approx 181$ K. Here we consider only one experimental condition, at the temperature of $T = 228$ K and at the q-vector of $0.63\ \mu m^{-1}$, where the long time behavior of the HD-TG signal shows a strongly nonexponential relaxation owing to the mixing of the structural and thermal modes. We will focus our attention on the ability of the hydrodynamic models of Sect. 3.2.2.1 to fit the long time tail of the data. A wide survey of the experimental result obtained on Glycerol can be found in [16, 17].

In Fig. 3.18, we show the HD-TG measurement. The signal shows very characteristic temporal regimes. In panel (a), we see a detail of the sound oscillations regime. In panel (b), when the sound attenuation is complete (i.e. the sound oscillations disappear), the sample enters in the isobaric regime; in this panel the signal is still changing due to the structural follow-up time. Finally, panel (c) shows the long times decay due both to the thermal diffusivity and the structural relaxation processes. The straight line drawn in the latter inset clearly shows that the signal tail needs a distribution of decaying exponential functions to be explained.

As already seen for the OTP case (cf. Sect. 3.4.2), the generalized viscoelastic model proposed in [36] is not able to reproduce the signal in this range of temperature. In particular, it fails to reproduce simultaneously the rise and the fall of the thermal signal at the temperatures where, owing to the mixing of structural and thermal modes, the fall becomes strongly non-exponential. As first step to resolve this issue, we applied the multiexponential model for a stretched exponential viscosity described at the end of Sect. 3.2.2. In Fig. 3.19, we report a fit of the data obtained with a distribution of 40 exponential functions. In the fit, most of the model parameters are fixed to the value reported in the literature. As can be seen, a satisfactory agreement is obtained in the fast part, as well as in the structural rise. As can be expected, the fitted sound speed jump $\Delta c_A^2 = c_\infty{}^2 - c_0{}^2$ is in good agreement with the data reported in literature [52].

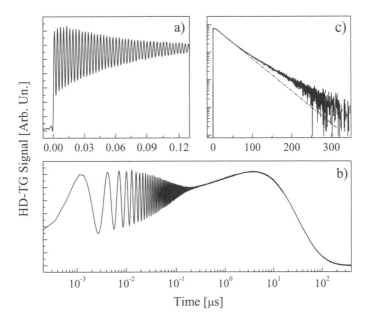

Fig. 3.18 HD-TG experimental time spectrum for glycerol in the supercooled phase ($T = 228°\mathrm{K}$) measured at $q = 0.63\,\mu\mathrm{m}^{-1}$ [16]. The panels show the various temporal region discussed in the text

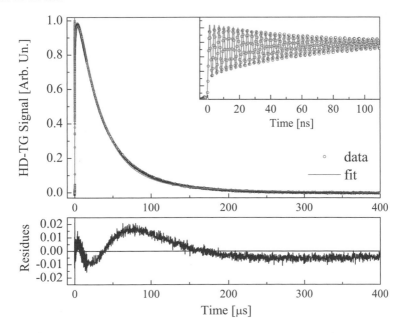

Fig. 3.19 Upper graph: Fit of the stretched-exponential multiexponential model (*gray line*) for the HD-TG data (*black dots*) of Fig. 3.18. Lower graph: The residuals of the fit are plotted on a magnified scale

We can observe a small but significant discrepancy in the long tail part, which can not be removed even with a larger number of exponentials. This part of the signal is determined by ΔD_v, which is in turn linked to the sound speed jump. A consequence of this wrong tail behavior is that the stretching parameter value obtained in the fit ($\beta = 0.34$) is not in good agreement with the value measured by light scattering.[11] A generalization of the model should include an additional memory term in the equations, without affecting the well-reproduced sound propagation part.

By following a phenomenological approach used in the 1980s to study the light scattering and forced light scattering data just on glycerol near to the glass transition [54,55], we add a relaxation behavior also to the thermal response. The equation to be generalized is the heat transport equation, allowing a temporal dependence in the diffusion coefficient D_t:

$$\delta \dot{T}_q + \frac{\gamma - 1}{\alpha} \psi_q - q^2 \gamma D_t \otimes \delta T_q = H. \tag{3.75}$$

Following the same steps applied for the the viscosity relaxation, we have

$$D_t(t) = D_t^{\infty} \delta(t) + \Delta D_t \sum_{i=1}^{m} \frac{G_t^{(i)}}{\tau_t^{(i)}} \exp(-\frac{t}{\tau_t^{(i)}}). \tag{3.76}$$

Also in this case the $G_t^{(i)}$ amplitudes are assumed to be normalized ($\sum_{i=1}^{m} G_t^{(i)} = 1$), and a δ-approximation is valid in the two extreme limits of a very slow and very fast responsive medium, the normalization constant ΔD_t being connected to the difference between the zero and the infinite frequency limits: $\Delta D_t = D_t(\omega = 0) - D_t^{\infty}$. Thus in these limits also for a time-dependent thermal diffusivity, the dynamics remain described by a matrix as in (3.56). As opposed to the viscosity case, now the D_t limits appear to be directly linked to the corresponding limiting values of the ratio $\lambda/\rho_0 C_p$. Neglecting the frequency dependence of the ratio λ/ρ_0 across the glass transition, we obtain a relation between the amplitude of the memory and the (inverse) specific heat jump: $\Delta D_t = (\frac{1}{C_p^0} - \frac{1}{C_p^{\infty}})\lambda/\rho_0$.

Defining again a set of auxiliary amplitudes

$$\zeta_t^{(i)}(t) = \frac{1}{\tau_t^{(i)}} \exp(-\frac{t}{\tau_t^{(i)}}) \otimes \delta T_q(t), \tag{3.77}$$

with the initial conditions $\zeta_t^{(i)}(t = 0) = 0$ and with the time evolution

$$\frac{\partial}{\partial t} \zeta_t^{(i)}(t) = -\frac{\zeta_t^{(i)}}{\tau_t^{(i)}} + \frac{\delta T_q}{\tau_t^{(i)}}, \tag{3.78}$$

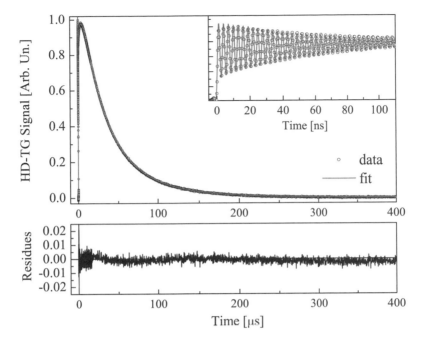

Fig. 3.20 Upper graph: The *gray line* is the fit for the generalized multiexponential model (stretched exponential in $D_v(t)$ simulated with $n = 40$ terms, plus a simple Debye term in $D_t(t)$, i.e. $m = 1$) for the HD-TG data of Fig. 3.18 (*black dots*). Lower graph: the fit residuals have been clearly reduced by the introduction of the D_t relaxation, as compared to Fig. 3.19

we obtain a new complete set of equations similar to the one reported in (3.56) whose solution can be obtained following (3.58). In Fig. 3.20, we show the result of the fit with just one exponential term added ($m = 1$). The fitted decay time is $\approx 24\,\mu s$, to be compared to the mean value $2.15\,\mu s$ of the KWW distribution. We see that the agreement between fit curve and data points is quite good in the whole temporal range. The fitted stretching parameter ($\beta = 0.39$) is in good agreement with the literature data. Moreover, the amplitude of the added term is comparable with the specific heat frequency jump estimation, $C_p^0 - C_p^\infty \approx 1\,\mathrm{J\,g^{-1}\,K^{-1}}$ [56].

Although it is based on conservation laws where opportune memory responses are introduced in the transport coefficients, the approach outlined here is a phenomenological one. An alternative point of view is offered by the non-equilibrium thermodynamics (NET) theory [57, 58]. In fact, NET provides an unified point of view to study the irreversible processes of heat conduction, diffusion, and viscosity. Promising results were obtained in the case of glycerol in the work of reference [16], where a simplified NET model was developed to obtain the isobaric evolution of the HD-TG signal. Further work is in progress to develop a NET model including the whole temporal range.

3.4.4   *Meta*-Toluidine: The Rotation-Translation Coupling Effect

As already stated at beginning of this section, the transient grating experiment with the selection of the polarization of the pump and probe beams enables us to highlight the presence of birefringence contributions. This experiment has been applied particularly to the study of the molecular glass former, *m*-toluidine, which shows a strong birefringence signal due to the rotation-translation coupling effect. The liquid has a melting temperature of 243 K and a glass transition temperature of 187 K.

In what follows we want to report only some HD-TG results obtained at $q = 1.00$ and $q = 0.630\,\mu m^{-1}$ from 298 K down to 180 K [23, 24]. The measurements were made at the following four configurations of polarization: VVVV, HHVV, VVHH, and HHHH. The data shown a relaxation dynamics scenario typical of a glass former with its acoustic, thermal, and structural modes. We shall not report the complete data analysis (from which the temperature and q behavior of acoustic, thermal, and structural relaxations have been extracted) but we will turn our attention particularly to the difference in signals corresponding to the different polarization configurations.

Whatever the temperature, the HD-TG signal does not show any detectable difference in profile or in intensity by changing the direction of polarization of the pump: $S_{VVVV} \simeq S_{VVHH}$ and $S_{HHHH} \simeq S_{HHVV}$ (see Fig. 3.21). This implies that the birefringence built directly by the pump electric field, the PIB effect, is not observed and can be neglected for the interpretation of the results. On the other hand, the TG signal profile does depend on the direction of

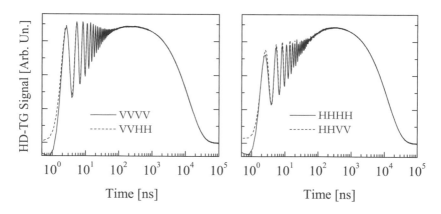

Fig. 3.21   Signals measured at $q = 1\,\mu m^{-1}$ and $T = 220$ K, for the four different polarization configurations. The S_{VVVV} and S_{VVHH} signals are shown in the left graph, the S_{HHHH} and S_{HHVV} ones in the right graph. Signals with identical probe and detection polarizations but different pump polarizations do not show any significant difference

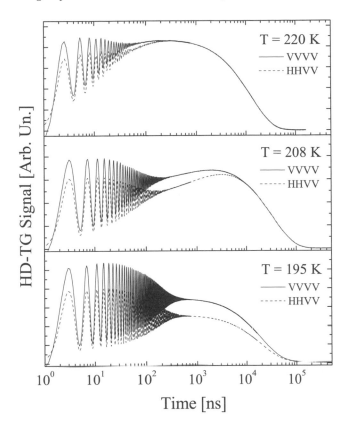

Fig. 3.22 Signals S_{VVVV} and S_{HHVV} collected on m-toluidine at $q = 0.630\,\mu m^{-1}$ at three different temperatures. As soon as the structural mode appears, the S_{VVVV} and S_{HHVV} data show a clear different shape of the signal profile in the short and intermediate time windows. At decreasing of temperature these differences extend over the whole temporal scale including the thermal decay

polarization of the probe, $S_{VVVV} \neq S_{HHVV}$. The difference starts below 230 K (at $q = 0.630\,\mu m^{-1}$), the temperature at which the structural relaxation begins to appear as a rise of the thermal contribution: the S_{HHVV} signal becomes less intense at short and intermediate times than the S_{VVVV} signal. At still lower temperatures, the difference extends to the whole time domain: the two signals differ only for an amplitude factor and they assume an identical temporal shape, Fig. 3.22. This birefringent signal must be ascribed to the R–TC effect. We recall that this effect arises from the coupling between the molecular velocity and the orientational variables. The heating and electrostriction produce initially a molecular velocity grating. In the points where the velocity gradient is maximum, the molecules tend to align along the motion direction giving rise to a birefringence grating. At high temperatures when the structural relaxation time is shorter than the acoustic wave period, this birefringent grating quickly relaxes, without affecting the measured dynamics, and the two signal S_{VVVV} and S_{HHVV} are the same. The difference between the two signals becomes clear as soon as the structural relaxation time exceeds the acoustic wave period. At these temperatures the signal depends also on the relaxation dynamics associated with the R–TC. The weight of this contribution is different for the two configurations of polarization VV and HH. At very low temperatures when the structural relaxation time is much longer than the thermal diffusion time, the grating of molecular alignment is practically frozen for the whole experimental time window and the two signals S_{VVVV} and S_{HHVV} differ only for an amplitude factor.

The interpretation of transient grating experiments involving birefringence effects requires the introduction of another thermodynamic variable, related to the mean molecular orientation, and the introduction of its motion equation. Furthermore, some coupling between the latter and the density must be introduced to account for R–TC effect. Recently Dreyfus et al. [32, 33], in order to analyze the light scattering spectra of metatoluidine, proposed a set of phenomenological equations to describe the equilibrium dynamics of liquids with axially symmetric molecules. Subsequently, some of these authors with other coworkers [23–27] applied this theoretical approach to analyze the transient grating experiment data collected on the same glass former. This hydrodynamic approach is extensively described in [30]. The theory starts from the idea that the dielectric constant change for a liquid formed of anisotropic molecules, has to be written as

$$\delta\epsilon_{ij}\left(\mathbf{q}, t\right) = a\delta\rho\left(\mathbf{q}, t\right)\delta_{ij} + bQ_{ij}\left(\mathbf{q}, t\right), \qquad (3.79)$$

where $a = \left(\partial\epsilon/\partial\rho\right)_T$ and Q_{ij} is a second-order symmetric traceless tensor characterizing the mean local orientation of the molecules [32, 33] (at thermal equilibrium all the Q_{ij} components are zero). Then, the hydrodynamics equations are rewritten in order to include a motion equation for the variables Q_{ij} and a coupling between $\psi = \nabla \cdot \mathbf{v}$ and Q_{ij}. The theory shows that, when the signal does not depend on the pump polarizations, pure density and rotational contributions can be extracted by performing two linear combinations of the S_{VVVV} and S_{HHVV} signals. In particular we have

$$S^{\mathrm{iso}} = \frac{1}{3}\left(2S_{\mathrm{VVVV}} + S_{\mathrm{HHVV}}\right) \qquad (3.80)$$

$$S^{\mathrm{aniso}} = \frac{1}{2}\left(S_{\mathrm{VVVV}} - S_{\mathrm{HHVV}}\right). \qquad (3.81)$$

Furthermore one deduces that the isotropic and anisotropic contributions can be related each other by the following expression:

$$S^{\text{aniso}}(q, t) \propto g(t) \otimes S^{\text{iso}}(q, t). \tag{3.82}$$

$g(t)$ is a function describing the relaxation process associated with the rotation-translation coupling. It is generally represented by a stretched exponential or its Fourier transform is represented by a Cole-Davidson function with a relaxation time τ_g and a stretching parameter β_g. Expressions 3.80 and 3.81 allow a simplification of the data analysis. In fact, since the isotropic part has been purged from the orientational dynamics, it can be analytically described, depending on the temperature range, by one of the hydrodynamic models reported in Sect. 3.2.2. One S_{iso} is parameterized, the fit of S_{aniso} requires only the determination of the two relaxation parameters τ_g and β_g. Here we report only two example of the isotropic and anisotropic contributions computed by using (3.80) and (3.81). These are shown in Fig. 3.23 for $T = 220\,\text{K}$ and $T = 208\,\text{K}$ at $q = 0.630\ \mu\text{m}^{-1}$. Quantitative fits of the isotropic and anisotropic parts of the data on the basis of the hydrodynamic model reported in [30] include the temperature behavior of all the interesting dynamics relaxation parameters, and are reported in references [23–25, 27].

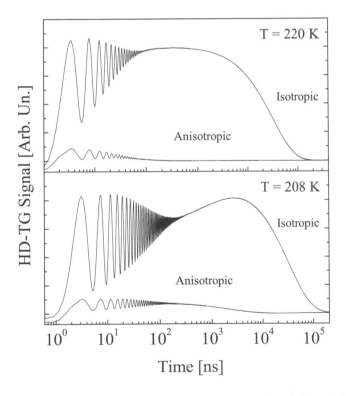

Fig. 3.23 The isotropic and anisotropic signals computed by using (3.80) and (3.81) for the TG data at 220 and 208 K

Acknowledgments

The authors express their acknowledgments to R.M. Pick for the crucial discussions and suggestions. We thank R. Righini for his continuous support. A special thanks is extended to J. Palmer for his critical reading. They further thank R. Ballerini, M. Giuntini, and M. De Pas, for their indispensable mechanical and electronic support. The present research has been performed at LENS and was supported by EC grant N.RII3-CT-2033-506350, CRS-INFM-Soft Matter (CNR), and MIUR-COFIN-2005 grant N. 2005023141-003.

Appendix: Mathematical Representation of the Stretched Exponential

In the case of glass-forming liquids, the experimental observations are generally described assuming a distribution function for the relaxation times contributing to each relaxation process $\phi(t)$ where, for example, $\phi(t) = D_{\rm v}(t)$. The approach is to adopt an empirical distribution function, $G_\alpha(\log \tau)$, whose shape is defined by a set of parameters α, defining a superposition of exponentially damped contributions, with time constant τ:

$$\phi(t) = \int_{-\infty}^{+\infty} \mathrm{d}(\log \tau) G_\alpha(\log \tau) \, \mathrm{e}^{-t/\tau}. \tag{3.A.1}$$

The parameters α are determined by a fit to the experimental data. We refer to the work of Bello et al. [59] for a clear summary of the various models and relative $G_\alpha(\log \tau)$. Here we focus on the stretched exponential function, and the relative Kohlrausch-Williams-Watts distribution:

$$\phi(t) = \mathrm{e}^{-(t/\tau_o)^\beta} = \phi_\beta(x), \tag{3.A.2}$$

$$\phi_\beta(x) = \mathrm{e}^{-x^\beta}, \tag{3.A.3}$$

where $x = t/\tau_o$. This distribution function has been used in a variety of experimental contexts, from dielectric and mechanical relaxation, light scattering, NMR, to optical Kerr effect measurements. Theoretical justifications of such a wide diffusion can be found in [46].

We will assume that the sum of exponentials in the expression (3.50) of the time-dependent viscosity is a stretched exponential. To be specific, following [59,60], the stretched exponential is written as superposition of Debye processes as

$$\phi_\beta(x) = \int_0^{+\infty} \mathrm{d}y \rho(y) \, \mathrm{e}^{-x/y}$$
$$= \int_{-\infty}^{+\infty} \mathrm{d}(\log y) G_\beta(\log y) \, \mathrm{e}^{-x/y}, \tag{3.A.4}$$

where $y = \tau/\tau_o$, and $G_\beta(\log y) = y\rho(y)$. The KWW distribution G_β is normalized (i.e. $\phi_\beta(0) = 1$) and the average relaxation time is given by

$$\langle \tau \rangle = \tau_o \int_{-\infty}^{+\infty} \mathrm{d}(\log y) G_\beta(\log y) y = \frac{\tau_o}{\beta} \Gamma\left(\frac{1}{\beta}\right).$$

For numerical calculation, the following integral expression of the KWW distribution is useful:

$$G_\beta(\log y) = \frac{1}{\pi y} \int_0^{+\infty} \mathrm{d}u \, \mathrm{e}^{-yu} \, \mathrm{e}^{-u^\beta \, \cos(\pi\beta)} \, \sin[u^\beta \, \sin(\pi\beta)].$$

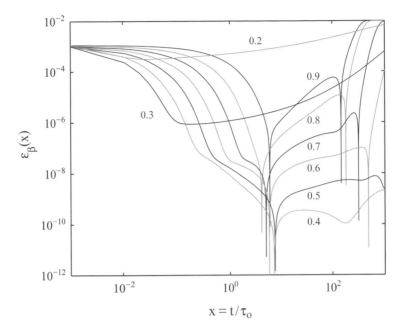

Fig. 3.A.1 Relative difference between the stretched exponential, $\phi_\beta(x)$, and the sum of n exponentials (for $n = 40$), $\phi_\beta^n(x)$ defined in (3.A.5), as a function of the (reduced) time

In fact, this integral can be calculated numerically by a nonadaptive double-exponential quadrature algorithm; the stretched exponential integral, (3.A.4), can thus be approximated by a sum of n exponentials evenly distributed on the logarithmic time scale:

$$\phi_\beta(x) \approx \phi_\beta^n(x) = \Delta \left(\log y\right) \sum_{i=1}^{n} G_\beta(\log y_i)\, e^{-x/y_i} \tag{3.A.5}$$

In Fig. 3.A.1, the residual $\epsilon_\beta = \left(\phi_\beta(x) - \phi_\beta^n(x)\right)/\phi_\beta(x)$ is reported, as a function of time, showing that, for $0.3 \leq \beta \leq 0.9$ and $n = 40$ we have a good reproduction: $\epsilon_\beta \leq 10^{-3}$ over more than four temporal decades.

Notes

1. Hellwarth developed the theoretical model starting from two hypotheses: weak electromagnetic fields and field wavelength much longer than the molecular dimensions. Thus he developed a perturbative theory in the dipole approximation.

2. Here we use the notation $\delta\epsilon$ instead of the $\delta\chi$, used in Chap. 2, in order to have a consistent notation with the quoted literature. We recall that when only the fluctuating or time-dependent part is relevant these two function are equivalent: $\delta\chi = \frac{1}{4\pi}\delta\epsilon$.

3. See for example Sect. 4.5 [30].

4. Indeed the general definition of the response function introduced in the Sect. 2 of Chap. 2 applies to the ISBS part of the TG response. Nevertheless, the calculations specific to the OKE response did not apply because these are restricted to the xyxy component and to the $q \to 0$ limit.

5. Contrary, in the water case the $(\partial\epsilon/\partial T)_\rho$ term can no longer be neglected [19–22].

6. Both of the forcing terms F and Q should contain another term dependent on $(\partial \epsilon / \partial T)_\rho$ [6, 31]. Nevertheless, these additional terms are always negligible compared to the usual electrostriction and heat absorption even when the contribution $(\partial \epsilon / \partial T)_\rho \delta T$ in (3.27) becomes important as, for instance, in the water case.

7. The analytical expressions are obtained by the hypotheses $D_v q^2 \ll c_0 q$, $\gamma q^2 D_T \ll c_0 q$. With these approximations, the roots of the determinant $M(s)$ can be easily calculated.

8. In most of cases, $\Gamma \approx D_v$ is a good approximation, even for weakly supercooled liquids, in the regime studied in the TG experiment.

9. In reality the average will be performed over the detector integration time, which even in ultrafast detectors is much longer than the time of optical cycle but generally shorter than the signal relaxation times.

10. Thanks to the small incidence angles of exciting fields, the polarization H is considered almost parallel to the q-direction of the induced grating.

11. If we take the β value from the work of [52] and we use the conversion formula between Cole-Davidson and KWW distribution [53], we obtain a literature reference value $\beta \approx 0.40$.

References

[1] Eichler H.J., Gunter P., Pohl D.W. (1986). *Laser-induced dynamic gratings*. Spriger-Verlag, Berlin.

[2] Eichler H.J., Salje G., Stahl H. (1973). Thermal diffusion measurements using spatially periodic temperature distributions induced by laser light, *J. Appl. Phys.* 44: 5383–5388.

[3] Pohl D.W., Schwarz S.E., Irniger V. (1973). Forced Rayleigh scattering, *Phys. Rev. Lett.* 31: 32–35.

[4] Bloembergen N. (1977). *Nonlinear optics*. Benjamin, New York.

[5] Hellwarth R.W. (1977). Third-order susceptibilities of liquids and solids, Part I of Vol.5 of Monographs: Progress in Quantum Electronics. Sanders J.H. and Stenholm S. (Eds.), Pergaman Press New York.

[6] Shen Y.R. (1984). *The principles of nonlinear optics*. Wiley, New York.

[7] Special issue on dynamic gratings and four-wave mixing, IEEE J. Quant. Electron. 22(8).

[8] Yan Y.X., Cheng L.T., Nelson K.A. (1988). The temperature-dependent distribution of relaxation times in glycerol: Time-domain light scattering study of acoustic and mountain-mode behavior in the 20 MHz–3 GHz frequency range, *J. Chem. Phys.* 88: 6477–6486.

[9] Duggal A.R., Nelson K.A. (1991). Picosecond–microsecond structural relaxation dynamics in polypropylene glycol: Impulsive stimulated light-scattering experiments, *J. Chem. Phys.* 94: 7677–7688.

[10] Silence S.M., Duggal A.R., Dhar L., Nelson K.A. (1992). Structural and orientational relaxation in supercooled liquid triphenylphosphite, *J. Chem. Phys.* 96: 5448–5459.

[11] Yan Y., Nelson K.A. (1987). Impulsive stimulated light scattering. I. General theory, *J. Chem. Phys.* 57: 6240–6256; ibid. Impulsive stimulated light scattering. II. Comparison to frequency domain light-scattering spectroscopy: 6257–6265.

[12] Yang Y., Nelson K.A. (1995). T_C of the mode coupling theory evaluated from impulsive stimulated light scattering on salol, *Phys. Rev. Lett.* 74: 4883–4886.

[13] Maznev A.A., Nelson K.A., Rogers J.A. (1998). Optical heterodyne detection of laser-induced grating. *Opt. Lett.* 23: 1319–1321.

[14] Goodno G.D., Dadusc G., Miller R.J.D. (1998). Ultrafast heterodyne-detected transient-grating spectroscopy using diffractive optics, *J. Opt. Soc. Am. B* 15: 1791–1794.

[15] Torre R., Taschin A., Sampoli M. (2001). Acoustic and relaxation processes in supercooled orthoterphenyl by optical-heterodyne transient grating experiment, *Phys. Rev. E* 64: 061504(1—10).

[16] Di Leonardo R., Taschin A., Sampoli M., Torre R., Ruocco G. (2003). Nonequilibrium thermodynamic description of the coupling between structural and entropic modes in supercooled liquids, *Phys. Rev. E (Rap. Com.)* 67: 015102-1–015102-4.

[17] Di Leonardo R., Taschin A., Sampoli M., Torre R., Ruocco G. (2003). Structural and entropic modes in supercooled liquids: experimental and theoretical investigation, *J. Phys. Condens. Matter* 15: S1181–S1192.

[18] Sampoli M., Taschin A., Eramo E. (2004). Hydrodynamic study of 3-methylpentane by transient grating experiment, *Philos. Mag.* 84: 1481–1490.

[19] Miller R.J.D. (1989). *Time resolved spectroscopy*. Clark R.J.H. and Hester R.E. (Eds.), Wiley, New York.

[20] Terazima M. (1996). Refractive index change by photothermal effect with a constant density detected as temperature grating in various fluids, *J. Chem. Phys.* 104: 4988–4998.

[21] Taschin A., Bartolini P., Ricci M., Torre R. (2004). Transient grating experiment on supercooled water, *Philos. Mag.* 84: 1471–1479.

[22] Taschin A., Bartolini P., Eramo R., Torre R. (2006). Supercooled water relaxation dynamics probed with heterodyne transient grating experiments, *Phys. Rev. E* 74: 031502(1–10).

[23] Taschin A., Torre R., Ricci M., Sampoli M., Dreyfus C., Pick R. (2001). Translation-rotation coupling in transient grating experiments: Theoretical and experimental evidences, *Europhys. Lett.* 56: 407–413.

[24] Pick R.M., Dreyfus C., Azzimani A., Taschin A., Ricci M., Torre R., Franosch T. (2003). Frequency and time resolved light scattering on longitudinal phonons in molecular supercooled liquids, *J. Phys. Condens. Matter* 15: S825–S834.

[25] Azzimani A., Dreyfus C., Pick R.M., Taschin A., Bartolini P., Torre R. (2007). A transient grating study of *m*-toluidine from 330 K to 190 K, *J. Phys. Condens. Matter* 19: 205146–205152.

[26] Azzimani A., Dreyfus C., Pick R.M., Bartolini P., Taschin A., Torre R. (2007). Analysis of a heterodyne-detected transient-grating experiment on a molecular supercooled liquid. I. Basic formulation of the problem, *Phys. Rev. E* 76:011509(1–6).

[27] Azzimani A., Dreyfus C., Pick R.M., Bartolini P., Taschin A., Torre R. (2007). Analysis of a heterodyne-detected transient-grating experiment on a molecular supercooled liquid. II. Application to m-toluidine, *Phys. Rev. E* 76:011510(1–10).

[28] Glorieux C., Nelson K.A., Hinze G., Fayer M.D. (2002). Thermal, structural, and orientational relaxation of supercooled salol studied by polarization-dependent impulsive stimulated scattering, *J. Chem. Phys.* 116: 3384–3395.

[29] Hinze G., Francis R.S., Fayer M.D. (1999). Translational-rotational coupling in supercooled liquids: Heterodyne detected induced molecular alignment, *J. Chem. Phys.* 111: 2710–2719.

[30] Pick R.M., Dreyfus C., Azzimani A., Gupta R., Torre R., Taschin A., Franosch T. (2004). Heterodyne detected transient gratings in supercooled molecular liquids: A phenomenological theory, *Euro. Phys. J. B* 39: 169–197.

[31] Franosch T., Pick R.M. (2005). Transient grating experiments on supercooled molecular liquids. II. Microscopic derivation of the phenomenological equations, *Eur. Phys. J. B* 47: 341–361.

[32] Dreyfus C., Aouadi A., Pick R.M., Berger T., Patkowski A., Steffen W. (1998). Light scattering measurement of shear viscosity in a fragile glass-forming liquid, metatoluidine, *Europhys. Lett.* 42: 55–60.

[33] Dreyfus C., Aouadi A., Pick R.M., Berger T., Patkowski A., Steffen W. (1999). Light scattering by transverse waves in supercooled liquids and application to metatoluidine, *Eur. Phys. J. B* 9: 401–419.

[34] Pick R.M., Franosch T., Latz A., Dreyfus C. (2002). Light scattering by longitudinal acoustic modes in molecular supercooled liquids. I. Phenomenological approach, *Eur. Phys. J. B* 31: 217–228.

[35] Franosch1 T., Latz A., Pick R.M. (2003). Light scattering by longitudinal acoustic modes in supercooled molecular liquids. II. Microscopic derivation of the phenomenological equations, *Eur. Phys. J. B* 31: 229–246.

[36] Yang Y., Nelson K.A. (1995). Impulsive stimulated light scattering from glass-forming liquids. I. Generalized hydrodynamics approach, *J. Chem. Phys.* 103: 7722–7731; ibid. Impulsive stimulated light scattering from glass-forming liquids. II. Salol relaxation dynamics, nonergodicity parameter, and testing of mode coupling theory, 103: 7732–7739.

[37] Berne B.B., Pecora R. (1976). *Dynamic light scattering*. Wiley, New York.

[38] Boon J.P., Yip S. (1980). *Molecular hydrodynamics*. McGraw-Hill, New York.

[39] Zwanzig R. (1965). Frequency-dependent transport coefficients in fluid mechanics, *J. Chem. Phys.* 43: 714–720.

[40] Mountain R.D. (1966). Thermal relaxation and brillouin scattering in liquids, *J. Res. Natl. Bur. Std.*, 70A: 207.

[41] Wong J., Angell C.A. (1976). *Glass structure by spectroscopy*. Marcel Dekker, New York.

[42] Debenedetti P.G. (1996). *Metastable liquids*. American Chemical Society, Washington.

[43] Fourkas J.T., Kivelson D., Mohanty U., Nelson K.A. (1996). *Supercooled liquids*. Princeton University Press, New Jersey.

[44] Götze W. (1999). Recent tests of the mode-coupling theory for glassy dynamics, *J. Phys. Condens. Matter* 11: A1–45.

[45] Cummins H.Z. (1999). The liquid-glass transition: A mode-coupling perspective, *J. Phys. Condens. Matter*, 11: A95–117.

[46] Angell A.C., Ngai K.L., McKenna G.B., McMillan P.F., Martin S.W. (2000). Relaxation in glassforming liquids and amorphous solids, *J. Appl. Phys.* 88: 3113–3157.

[47] Terazima M. (1999). Optical heterodyne detected transient grating for the separations of phase and amplitude gratings and of different chemical species, *J. Phys. Chem. A* 103: 7401–7407.

[48] Mukamel S. (1995). *Principles of nonlinear optical spectroscopy*. Oxford University Press, New York.

[49] Winkler K., Lindner J., Vöhringer P. (2002). Low frequency depolarized Raman-spectral density of liquid water from femtosecond optical Kerr-effect measurements: Lineshape analysis of restricted translational modes, *Phys. Chem. Chem. Phys.* 4: 2144–2155.

[50] Torre R., Bartolini P., Righini R. (2004). Structural relaxation in supercooled water by time-resolved spectroscopy, *Nature* 428: 296–299.

[51] Angell C.A. (1988). Perspective on the glass transition, *J. Phys. Chem. Solids* 49: 863–871.

[52] Comez L., Fioretto D., Scarponi F., Monaco G. (2003). Density fluctuations in the intermediate glass-former glycerol: A Brillouin light scattering study, *J. Chem. Phys.* 119: 6032–6043.

[53] Lindsey C.P., Patterson G.D. (1980). Detailed comparison of the Williams-Watts and Cole-Davidson functions, *J. Chem. Phys.* 73: 3348–3357.

[54] Lin Y.-H., Wang C.H. (1979). Rayleigh–Brillouin scattering and structural relaxation of a viscoelastic liquid, *J. Chem. Phys.* 70: 681–688.

[55] Allain C., Berard M., Lallemand P. (1980). Thermal diffusivity of glycerol at the liquid-glass transition, *Molec. Phys.* 41: 429–440.

[56] Rajeswari M., Raychaudhuri A.K. (1993). Specific-heat measurements during cooling through the glass-transition region, *Phys. Rev. B* 47: 3036–3046.

[57] Allain C., Lallemand P. (1979). Phenomenological models compatible with acoustical and thermal properties of viscous liquids, *J. Phys* 40: 679–692.

[58] De Groot S.R., Mazur P. (1969). *Non-equilibrium thermodynamics*. North-Holland, Amsterdam.

[59] Bello A., Laredo E., Grimau M. (1999). Distribution of relaxation times from dielectric spectroscopy using Monte Carlo simulated annealing: Application to a-PVDF, *Phys. Rev. B* 60: 12764–12774.

[60] Alvarez F., Alegra A., Colmenero J. (1991). Relationship between the time-domain Kohlrausch-Williams-Watts and frequency-domain Havriliak-Negami relaxation functions, *Phys. Rev. B* 44: 7306–7312.

Chapter 4

DYNAMICAL PROCESSES IN CONFINED LIQUID CRYSTALS

M. Vilfan, I. Drevenšek Olenik, and M. Čopič

Abstract   In liquid crystals, incident light is strongly scattered due to coupling of light with orientational director fluctuations. A brief introduction to the orientational dynamics in bulk systems will be given, followed by several examples of the dynamics in confined samples, in detail discussed for thin slabs and for ellipsoidal droplets in holographic polymer dispersed liquid crystals. The influence of the confining surface on the scattered light will be shown and the method with which the information about the interaction with the surface has been obtained will be described. We will show that dynamic light scattering can be successfully used for measurements of surface anchoring coefficients and to probe the anchoring transitions in various confined geometries.

4.1   Introduction

Dynamic light scattering refers to the scattering of light due to time-dependent fluctuations of the optical dielectric function in a medium on time scales of the order of nanoseconds or slower. Nowadays it is usually analyzed with the technique of photon correlation spectroscopy and is a very useful tool to study the dynamics of fluctuations in soft matter physics. In this chapter we will show how it can be used to investigate dynamical properties of liquid crystals confined to finite geometries with dimensions comparable to the wavelength of light or smaller. In such samples, the dynamics of fluctuations observed by light scattering can become strongly influenced by the interactions of liquid crystal with the surface. In addition, because of the finite system size, the momentum of light and of medium excitations need no longer be conserved in scattering, and this profoundly affects the relation between the spectrum of the scattered light and the fluctuations in the liquid crystalline sample.

Liquid crystals are materials that shows some characteristics typical for crystalline phases despite their fluidity. This is possible due to anisotropic molecules that exhibit orientational and no – or only partial – positional order. The simplest liquid crystalline phase is the nematic phase, with no translational ordering at all. In this case, the elongated molecules are locally oriented along a particular direction – the director. A higher degree of order is found at lower temperatures in the smectic phases, where the oriented molecules form fluid layers. The director can be either perpendicular to the smectic layers (smectic A phase) or tilted with respect to the layer normal (smectic C and other tilted phases). Because of their optical and mechanical properties, based on their partial ordering, liquid crystals have been found very useful for electrooptic applications such as displays and optical switches. Besides the industrial importance, liquid crystalline materials are interesting also as model systems for studying a broad spectrum of fundamental phenomena in physics due to the richness of different phases and structures. For instance, in liquid crystalline phases various finite size effects are very profound, hence the associated modifications of the structures and phase transitions can be studied in details. For a good introduction to the fundamentals of liquid crystals see [1, 2].

Here, we focus only on the nematic phase and describe light scattering by the fluctuations of the nematic director. In the nematic phase, the orientational ordering of the molecules results in a local optical uniaxiallity, with the optic axis parallel to the director. Because of the fluid nature of the phase, the orientation of the director can vary through the sample and be thus a function of position. As will be discussed in the following sections, there is an increase in the elastic energy associated with such deformations. It is, however, fairly small so that for example in a millimeter-sized sample – and in the absence of external fields – the director will in general be disordered. External influences, such as electric or magnetic field, can easily change the orientation of the director and align the sample. This reorientation process, together with large optical anisotropy, is the basis for successful application of nematic liquid crystals in display devices.

The orientation of the director is not constant in time and fluctuates due to thermal excitations. Because the energy associated with the local perturbation of the director is small, the amplitude of the thermal fluctuations is large. With the director, also the optic axis fluctuates and with it the optical dielectric constant, which results in strong scattering of light. This feature is most prominently observed in the opaque appearance of liquid crystals.

The dynamics of orientational director fluctuations is governed by the elastic and viscous properties of the liquid crystal: local perturbations of the director are driven back to equilibrium by elastic torques that are opposed by a viscous drag on molecular rotation. This viscous drag is fairly large, making inertial effects negligible, and the orientational fluctuations overdamped. Both viscous and elastic coefficients – which are among the most important parameters for quantitative description of a liquid crystal – have often been measured using light scattering experiments [3, 4].

In recent years, there has been a considerable interest in not only bulk properties of liquid crystals, such as viscoelastic coefficients, but also in liquid crystals confined to small cavities of different geometries. A detailed description is found in [5]. Liquid crystals in such systems are very appropriate for observing the effects of confinement and disorder on phase transitions, and for studying the influence of confining surfaces on static and dynamic properties.

As will be shown, the confinement changes the behaviour of fluctuations and associated light scattering in several important ways. In a finite geometry, for example, the spectrum of fluctuations is no longer continuous but becomes discrete. When liquid crystal is embedded in many small cavities, the surface to volume ratio becomes large and thus interactions of liquid crystal with surfaces become important or even dominant. In some special cases, this allowed us to obtain parameters describing surface interactions via light scattering, as will be described in detail in the following sections [6–9].

After an overview of the continuum description of nematic liquid crystals, the scattering of light by the director fluctuations is presented. Some examples of confined liquid crystal geometry are described and the effect of the confinement on the fluctuation modes and on the scattered light is discussed for two different geometries. Experimental observations show that the light scattering experiment can be used for determining the degree of interaction of liquid crystal with the confining surface and to analyse a critical slowing down of fluctuations in the vicinity of structural transitions.

4.2 Continuum Description of Nematic Liquid Crystals

The theoretical description of the nematic liquid crystals is based on the macroscopic theory of elasticity. It introduces an order parameter that is zero in the isotropic phase and nonzero in the nematic, and can be directly related to macroscopic quantities, such as the anisotropy of the magnetic susceptibility. Because of the symmetry properties of the uniaxial nematic phase, the order parameter Q is a traceless tensor. Defining the director $\mathbf{n}$ as a unit vector parallel to the local average orientation of the elongated molecules, the tensor order parameter Q can be written as

$$Q_{\alpha\beta} = S \left(\frac{3n_\alpha n_\beta - \delta_{\alpha\beta}}{2} \right). \tag{4.1}$$

The parameter S is introduced as the scalar order parameter. It measures the degree to which the molecules are ordered and aligned parallel to the director. At high temperatures, the sample is in the isotropic phase and $S = 0$. When the temperature is lowered and the transition point from the isotropic to the nematic phase is reached, the scalar order parameter abruptly changes to $S \approx 0.4$ and continuously increases with decreasing temperature to $S \approx 0.6$ [1].

Other macroscopic properties can also be related to the tensor order parameter – in fact, any second rank tensor property of the system can be expressed in terms of Q. Using (4.1), the dielectric tensor can be written as [1]

$$\varepsilon_{\alpha\beta} = \varepsilon_\perp \delta_{\alpha\beta} + (\varepsilon_\| - \varepsilon_\perp) n_\alpha n_\beta, \qquad (4.2)$$

where the subscripts $\|$ and $\perp$ are defined relative to the director $\mathbf{n}$ and the anisotropy $\varepsilon_a = (\varepsilon_\| - \varepsilon_\perp)$ is proportional to the scalar order parameter S.

When theoretically considering nematic liquid crystals, it is often assumed that the scalar order parameter S is constant through the sample. This is indeed true if the gradients in the sample are small. In this case, the free energy of a bulk liquid crystal can be obtained by calculating the elastic energy associated to the spatial variations in $\mathbf{n}$. Neglecting the surface terms, the well-known Frank elastic energy is obtained:

$$F_{el} = \frac{1}{2} \int \left[K_1 \left(\nabla \cdot \mathbf{n}\right)^2 + K_2 \left(\mathbf{n} \cdot (\nabla \times \mathbf{n})\right)^2 + K_3 \left(\mathbf{n} \times (\nabla \times \mathbf{n})\right)^2 \right] dV.$$
$$(4.3)$$

The constants $K_{i=1-3}$ are the Frank elastic constants and correspond to the three fundamental deformations of the director field: splay, twist and bend. Their magnitude is of the order of 10^{-11} N, which is a fairly small value, meaning that on a length scale of several hundred microns (a typical size of a bulk sample), the nematic liquid crystal can easily deform or be deformed. It should be added that the Frank elastic constants depend on the order in the liquid crystal, being approximately proportional to S^2. Consequently, the Frank constants K_i are temperature dependent and increase with decreasing temperature.

If additional external fields act on the director, e.g. the electric field, or in the vicinity of a confining surface, additional terms need to be considered in (4.3). By applying electric field E, the free energy is changed by $-\frac{1}{2}\varepsilon_a(\mathbf{E} \cdot \mathbf{n})^2$. For positive dielectric anisotropy, i.e. $\varepsilon_\| > \varepsilon_\perp$, which is often the case, this causes the director to orient parallel to the electric field.

The contribution of the confining surface to the free energy depends strongly on the characteristics of the substrate. In the case where the surface induces a preferred orientation ν of the director at the boundary, the intrinsic interactions orient the whole sample and the sample is aligned. Any deviation of the director at the surface from the induced orientation increases the free energy, and the term added to (4.3) is the Rapini-Papoular surface anchoring term [10]

$$F_{RP} = -\frac{1}{2}W(\mathbf{n} \cdot \nu)^2 \qquad (4.4)$$

The quantity W is introduced as the surface anchoring energy coefficient and its typical value is of the order of $10^{-5} - 10^{-4}$ J m^{-2}. Often the ratio $\lambda = K/W$ is introduced as the extrapolation length, with values of the order of $10^{-6} - 10^{-5}$ m.

By minimising the complete free energy with respect to the variations in the director $\mathbf{n}$, which preserve the condition $|\mathbf{n}| = 1$, the equilibrium static configuration of the nematic system can be obtained. However, there are several dynamic processes present in a nematic liquid crystal, which will be discussed in the following section.

4.3   Fluctuational Dynamics in Nematic Liquid Crystals

Dynamic processes that are specific for liquid crystalline phases are the spatially correlated collective orientational fluctuations. These collective fluctuations include the director fluctuations, describing the local fluctuations in the orientation of the director, and the order parameter fluctuations, describing the fluctuations in the magnitude of S. However, in most cases (except in the vicinity of a phase transition), the latter can be neglected and the orientational director fluctuations are dominant.

To obtain the relaxational dynamics of the director fluctuations, the equation of motion for nematics needs to be written, taking into account that once a deformation or distortion in the director field is excited, the director returns into its equilibrium configuration. The elastic torque is balanced by the viscous torque, and the following relation is obtained [1]:

$$\eta \frac{\partial \mathbf{n}(\mathbf{r}, t)}{\partial t} = K \left[\nabla^2 \mathbf{n} - \left(\mathbf{n} \cdot \nabla^2 \mathbf{n} \right) \mathbf{n} \right] \qquad (4.5)$$

Equation 4.5 is written in the one-constant approximation ($K_i = K$) for simplicity reasons and the parameter η is the effective rotational viscosity. In principle, in an incompressible nematic there are five independent viscosity coefficients, three for describing the dissipative stress due to the fluid velocity gradient, one coupling the director with flow and one due to the rotation of the director with respect to the fluid. However, the fluid flow can often be neglected and only an effective rotational viscosity η is used.

Equation of motion (4.5) can be used to study the thermally excited director fluctuations, with the director being a sum of an equilibrium and a fluctuating part

$$\mathbf{n}(\mathbf{r}, t) = \mathbf{n}_0 + \delta \mathbf{n}(\mathbf{r}, t). \qquad (4.6)$$

In the first approximation, the fluctuating part is perpendicular to the equilibrium director and its size far from the phase transitions is small: $\delta \mathbf{n} \ll 1$. Equation of motion can thus be linearised and solved.

In an infinite sample, the solutions for the director fluctuations $\delta \mathbf{n}$ have the form of plane waves with a wave vector $\mathbf{q}$. The wave vectors are in principle arbitrary, limited only by the sample dimensions on one side and with the requirements of the continuum approach on the other, and the spectrum of $\mathbf{q}$ is continuous. In such system, two purely dissipative eigenmodes can be identified for each $\mathbf{q}$: the first one is a combination of splay and bend fluctuations and the second one a combination of twist and bend fluctuations. The amplitude of each of these two eigenmodes ($\alpha = 1, 2$) with a given wave vector $\mathbf{q}$ can be calculated using the equipartition theorem

$$\left\langle |\delta \mathbf{n}_\alpha (\mathbf{q})|^2 \right\rangle = \frac{1}{V} \frac{k_B T}{K_\alpha q_\perp^2 + K_3 q_\parallel^2}. \tag{4.7}$$

Here, k_B is the Boltzmann constant, T the temperature, V the volume of the sample and the components of the wave vector, $q_\perp$ and $q_\parallel$, are written relative to the equilibrium director orientation $\mathbf{n}_0$. From (4.7) it can be seen that the amplitude of the fluctuations with the smallest wave vector is the largest.

Inserting (4.6) into the equation of motion (4.5), linearising for small fluctuations, and solving the dynamical equation for incompressible nematics, the relaxation rates for the two eigenmodes are

$$\frac{1}{\tau_\alpha} = \frac{K_\alpha q_\perp^2 + K_3 q_\parallel^2}{\eta_\alpha(\mathbf{q})} \approx \frac{K}{\eta} q^2. \tag{4.8}$$

The viscosities η_α are different for the two fluctuation modes and even depend on the orientation of $\mathbf{q}$ with respect to $\mathbf{n}$. However, the corrections are small and η_α are often replaced with an effective rotational viscosity η. The relaxation rates can be further simplified if the one-constant approximation is used.

These two low-frequency dissipative modes have also been experimentally observed and the theoretically predicted dispersion relations in bulk nematic liquid crystals (4.8) were confirmed [3, 4].

4.4   Light Scattering by Director Fluctuations

Probably the most evident consequence of the director fluctuations is the opaque appearance of nematic liquid crystals. This is because the optical axis coincides with the director and as the director fluctuates, so does the dielectric tensor and any light hitting the sample is scattered. Using (4.2), the relation between the fluctuations in the dielectric tensor and director fluctuations can be obtained:

$$\delta\varepsilon = \varepsilon_a (\mathbf{n} \otimes \delta\mathbf{n} + \delta\mathbf{n} \otimes \mathbf{n}) \tag{4.9}$$

Fluctuations in the dielectric tensor $\delta\varepsilon$ scatter light and the total scattered field is written as the sum of dipole radiation induced by the incoming light field over the scattering volume [1]:

$$E_s (\mathbf{q}_s, t) \propto \int_{V_s} e^{i\mathbf{q}_s \cdot \mathbf{r}} (\mathbf{f} \cdot \delta\varepsilon (\mathbf{r}, t) \cdot \mathbf{i}) \, dV, \tag{4.10}$$

where the scattering vector $\mathbf{q}_s$ is introduced as the difference between the wave vectors of the incident $\mathbf{k}_i$ and scattered wave $\mathbf{k}_f$. The scattered electric field E_s

depends strongly on the polarisations of the initial **i** and scattered wave **f**. A proper selection of the polarisations of the incoming and scattered light beams allows separate observations of the two orientational fluctuation modes.

Usually the dimensions of the sample and the scattering volume V_s are much larger than $1/q_s$. In this case the integral in (4.10) can be simplified and only the Fourier component of the fluctuations with wave vector equal to the scattering vector contributes to scattering of light: $E_s \propto (\mathbf{f} \cdot \delta\varepsilon(\mathbf{q}_s, t) \cdot \mathbf{i})$. If the sample is fairly thin or small, several contributions need to be taken into account, and this will be discussed in the following sections.

The amplitude of the scattered field E_s varies with time due to the fluctuations in the director orientation and hence the dielectric tensor. Therefore, a very convenient way to observe the director dynamics is to study the dynamics of the scattered light, i.e. to use dynamic light scattering for measuring the autocorrelation function of the scattered light. The normalised intensity autocorrelation function $g^2(\mathbf{q}, t')$ is defined as [11]

$$g^{(2)}(\mathbf{q}, t') = \frac{\langle I(\mathbf{q}, t) I(\mathbf{q}, t + t') \rangle}{\langle I(\mathbf{q}, t) \rangle^2}, \qquad (4.11)$$

where $I(\mathbf{q}, t)$ is the intensity of the scattered light, proportional to E_s^2. In general, two different regimes of measurement can be considered: homodyne and heterodyne regime. In the former case, the intensity of the scattered light by the fluctuations is much larger than the intensity of the statically scattered light in the system, and the dynamically scattered light interferes at the detector. In the latter case, which is valid in samples with a small amount of the scattering medium, the contribution of the dynamically scattered light is small compared to the statically scattered light, and as the two contributions are statistically independent, the autocorrelation function is

$$\left(g^{(2)}(\mathbf{q}, t') - 1 \right) \propto \langle E_s^*(\mathbf{q}, t) E_s(\mathbf{q}, t + t') \rangle \propto g^{(1)}(\mathbf{q}, t'). \qquad (4.12)$$

It can be shown that this is further proportional to $\langle \delta\mathbf{n}(t) \, \delta\mathbf{n}(t + t') \rangle$. The measured autocorrelation function of the scattered light thus directly yields information about the orientational fluctuations at a given scattering vector $\mathbf{q}$. In heterodyne regime, the decay time of $g^{(2)}(\mathbf{q}, t')$ equals the relaxation time of the director fluctuations [11].

Dynamic light scattering has often been used in bulk liquid crystals for several different purposes. In most cases, the ratios of the viscoelastic coefficients were determined, according to the dispersion relation (4.8). In principle, from the amplitude of the non-normalised autocorrelation function one can also obtain the elastic constants themselves by using (4.7), but as it is difficult to measure absolute intensities in light scattering, Frank elastic constants are usually obtained using other methods.

Recently, dynamic light scattering has also been widely used to study the dynamics in confined liquid crystalline systems and a brief overview of different confinement geometries will be given in the following section.

4.5   Confined Liquid Crystals

Confined liquid crystalline samples are systems in which the confining boundaries strongly influence both the static and the dynamic properties of the liquid crystal. Depending on the type and the shape of the surface, the order in a liquid crystal can be either enhanced or reduced, liquid crystal molecules can be in average oriented in one direction or otherwise manipulated. Confinement geometries usually include thin planar slabs, cylindrical pores (e.g. nuclepore membranes), liquid crystalline droplets or random structures in porous glasses and silica aerogels. For an extensive review see [5]. Here, only some of the most important confinement geometries will be presented, including planar cells and two examples of spherical droplets.

4.5.1   Planar Cells

Planar confinement is by far the simplest and most common case of liquid crystal confinement as it is used in the LC displays. In such samples, liquid crystal is placed between two solid flat surfaces. Additional influence of the surface is achieved by applying an aligning layer to the boundaries, which induces a director orientation and results in an aligned liquid crystal cell. The preferred orientation can be either in the plane of the surface (planar alignment), perpendicular to it (homeotropic alignment) or tilted with respect to the boundary. Further classification of cells can be made when comparing the orientation, which is induced on the two confining boundaries: if both surfaces induce the same director orientation, the sample is either homogeneously planar or homeotropic. If two plates inducing planar alignment are rotated with respect to each other, twisted planar structure is obtained. Because of the symmetry of the nematic director, a maximum angle of $90°$ is possible. This is the geometry used in the display devices. The third class of planar cells is achieved by using surfaces, where one surface induces planar and the other homeotropic alignment, resulting in a homogeneous bend deformation throughout the sample.

When studying the dynamics in the sample, usually the simplest geometry is used – either homogeneous planar or homeotropic alignment. This enables direct measurements of the interactions between liquid crystals and aligning surfaces as well as investigations how a confining boundary affects the liquid crystal. To make the observed effects as strong as possible, the thickness of a cell should not exceed a few microns. It is also clear that the thinner the cell is, the more pronounced are the effects of confinement.

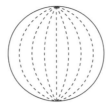

Fig. 4.1 Bipolar structure of the nematic director in a droplet where planar anchoring is induced by the polymer matrix

4.5.2 Polymer Dispersed Liquid Crystals

A more complex system of confined liquid crystals are polymer dispersed liquid crystals (PDLCs). These systems are usually obtained by mixing a liquid crystal with a suitable monomer. The monomer is then polymerised resulting in the separation of the liquid crystal and the polymer. In this manner, micrometer-sized droplets of liquid crystal are formed in a polymer matrix [12]. By varying the polymerisation and phase separation conditions, the size of the droplets can be controlled.

In spherical confinement, a number of different structures can exist, and their appearance strongly depends on the boundary conditions at the surface [13]. In the case of planar anchoring, the most common director configuration is the bipolar structure (Fig. 4.1), where because of the spherical geometry, two singular points appear. Another important feature of the bipolar structure in nematic droplets is a net average orientation of the molecules. However, the director orientation in different droplets is randomly distributed and any incident light is strongly scattered. By applying an electric field, the bipolar axes of the droplets orient perpendicular to the PDLC film and by matching the index of refraction of the polymer with the ordinary refractive index of the liquid crystal, the film becomes transparent. Electric field thus easily switches a PDLC film from opaque to transparent state, a process that is used for manufacturing large switchable windows at low costs.

4.5.3 Holographic Polymer Dispersed Liquid Crystals

Holographic polymer dispersed liquid crystals (H-PDLCs) are obtained by photopolymerisation of a liquid crystal–prepolymer mixture in the interference field of two or more laser beams [14, 15]. This induces phase separation in which liquid crystalline material predominantly congregates in dark regions of the optical interference pattern [16, 17]. As a result, holographic gratings with extremely high refractive-index contrast are formed. In addition, the diffraction efficiency of these gratings is electrically switchable; the application of an electric field results in reorientation of the nematic director field within

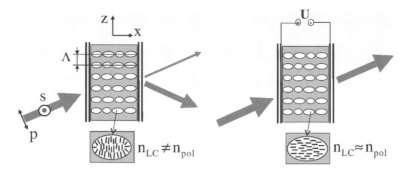

Fig. 4.2 The structure and operation of the H-PDLC transmission gratings. Liquid crystal droplet planes are perpendicular to the glass plates. The grating spacing Λ is typically in the range of $0.5 - 5$ μm. In the absence of external voltage the droplet directors are preferentially aligned perpendicular to the droplet planes, while above some threshold voltage they reorient along the electric field, i.e. perpendicular to the glass walls. This causes a decrease of the refractive index mismatch between the liquid crystal domains and the polymer matrix and hence the diffraction efficiency is reduced. The directions of s and p polarisations of incident optical beam are also denoted

the liquid crystal domains and similar as in PDLCs, the refractive index mismatch between the liquid crystal regions and the polymer network is modified (Fig. 4.2) [18]. These features make H-PDLCs very promising media for potential applications in switchable holographic elements, as well as for various kinds of reflective display devices.

In contrast to PDLC structures, H-PDLC media exhibit a strong inherent ordering of droplet directors, which is related to strongly elongated shape of the liquid crystal domains. The preferential orientation of the nematic director field is usually perpendicular to the liquid crystal rich planes [16, 19, 20]. For that reason, in the absence of external electric field, H-PDLCs exhibit much less static light scattering than PDLCs. Another consequence is that their diffraction efficiency strongly depends on the polarisation state of the incident light and is usually much larger for polarisation perpendicular to the holographic planes (p-polarisation) than for polarisation along the holographic planes (s-polarisation). This is observed even for structures that exhibit irregular nondroplet morphologies, in which the alignment of liquid crystal domains is attributed to a scaffolding effect of the polymer network [21].

4.6 Fluctuation Spectrum in Confined Geometries

Dynamic properties of a liquid crystal and spectrum of orientational director fluctuations, discussed previously for bulk samples, are strongly influenced by the presence of a surface. As the system is no longer infinite and homogeneous, the fluctuation modes are no longer plane waves with an arbitrary wave vector **q**.

Instead, the director fluctuation eigenmodes are at least in one direction standing waves, depending on the geometry of the confinement. In planar cells, the eigenmodes in the plane of the confinement are still propagating waves, whereas in the direction perpendicular to the boundaries, they are sinusoidal standing waves [22]. In cylindrical geometry, the eigenwaves are similar to Bessel and Neumann functions [23] and in the case of nematic droplets, they resemble spherical Bessel functions [24]. Because of geometry limitations, the set of allowed eigenwave vectors **q** is discrete and the wave vector values depend on the system geometry, size of the sample and on the boundary conditions. A detailed analysis will be given for planar and spherical geometry but let us first discuss the problem in general.

Similar as for infinite samples, the dynamical properties and behaviour can be obtained if the hydrodynamic equation (4.5) is considered. Linearising for small director fluctuations $\delta\mathbf{n}$ and taking into account the pure relaxational nature of the fluctuating modes (4.8), the following relation is obtained:

$$\nabla^2\delta\mathbf{n} - \left(\mathbf{n}_0 \cdot \nabla^2\mathbf{n}_0\right)\delta\mathbf{n} - \left(\mathbf{n}_0 \cdot \nabla^2\delta\mathbf{n}\right)\mathbf{n}_0 - \left(\delta\mathbf{n} \cdot \nabla^2\mathbf{n}_0\right)\mathbf{n}_0 = -q_N^2\delta\mathbf{n},$$
(4.13)

where q_N are the discrete eigenvalues of the fluctuation modes, determined by the boundary conditions. The boundary conditions are obtained by minimising the free energy at the boundary with respect to the director orientation. If ν is the orientation of the director induced by the surface (the easy axis), the boundary conditions are [22]

$$K\left(\sigma \cdot \nabla\right)\delta\mathbf{n} + W\left(2\left(\nu \cdot \mathbf{n}_0\right)\left(\nu \cdot \delta\mathbf{n}\right)\mathbf{n}_0 - \left(\nu \cdot \delta\mathbf{n}\right)\nu + \left(\nu \cdot \mathbf{n}_0\right)^2\delta\mathbf{n}\right)$$
$$= -\zeta\frac{\partial\delta\mathbf{n}}{\partial t}.$$
(4.14)

Parameter W is the surface anchoring strength as introduced in (4.4), σ is unit vector normal to the surface and ζ is the surface orientational viscosity. In most cases, the surface viscosity is very small and can be neglected. However, in particular confinement geometries, such as cylindrical cavities, the effect of surface viscosity on the dynamics has been observed and the corresponding characteristic length was found to be of the order of 10 nm [6]. The origin of the surface viscosity, however, is still not very clear. It can be either due to dissipation resulting from surface processes like adsorption–desorption, or slipping of the molecules on the confining surface.

4.6.1 Fluctuation Dynamics in Thin Planar Cells

Relaxation dynamics in thin planar cells, i.e. in systems where the liquid crystal is confined only in one direction, has been extensively studied both theoretically [22] and experimentally [25]. It has mostly been used to study interactions of liquid crystal with boundaries and aligning layers.

Let us assume the sample is a uniform nematic layer placed between two equally treated glass plates. Then the eigenmodes of the orientational fluctuations are standing waves of sinusoidal shape. Defining the z axis as the one perpendicular to the nematic layer and choosing the origin $z = 0$ to be in the centre between the glass plates, $\cos(q_{zN}z)$ and $\sin(q_{zN}z)$ are the solutions for even and odd fluctuation eigenmodes, respectively, where q_{zN} are the components of the eigenvalues in the direction of confinement. The eigenvalues q_{zN} are given by secular equations, which are obtained from the boundary conditions (4.14)

$$ q_{zN} \tan\left(\frac{q_{zN}d}{2}\right) = \frac{W}{K} \quad \text{and} \quad \frac{1}{q_{zN}} \tan\left(\frac{q_{zN}d}{2}\right) = -\frac{K}{W}, \qquad (4.15) $$

for even and odd fluctuation eigenmodes, respectively, where d is the thickness of the nematic layer. The boundary conditions are valid for both fluctuation modes if proper elastic constants and anchoring coefficients are taken into account: for fluctuations of the director from the easy axis in the plane parallel to the boundaries, the corresponding elastic constant is K_2, and W_φ is the azimuthal (in-plane) anchoring coefficient. If fluctuations out of the boundary plane are considered, K is the splay constant K_1 and W_ϑ the zenithal (out-of-plane) anchoring coefficient.

The secular equations, which define the fluctuation eigenvectors, in general cannot be solved analytically. Therefore often two limiting cases are considered: the first one is valid when the orientational interaction with the surface – the anchoring – is weak and the extrapolation length $\lambda = K/W$ large compared to the sample thickness d, and the second regime when the extrapolation length is small with respect to d.

Infinitely strong anchoring at the boundaries means that the director orientation at the surface does not deviate from the induced orientation (easy axis) and the amplitude of the fluctuations at the surface equals 0. The secular equations in this limiting case can be solved and the eigenvalues are $q_{zN} = N\pi/d$, where ($N = 1, 2, ...$). In this case the fluctuation relaxation time depends only on the bulk viscoelastic properties of liquid crystal and the thickness of the layer d:

$$ \tau_N = \frac{\eta}{K}\frac{d^2}{N^2\pi^2}. \qquad (4.16) $$

If anchoring at the boundaries is strong but finite, the extrapolation length is small compared to the inverse fluctuation wave vector component and thus to d. The secular equation can be expanded in terms of small deviations in the wave vector from the value for infinite anchoring and for the fundamental mode ($N = 1$) yields [8]

$$ \tau_0 \approx \frac{\eta}{K}\frac{d^2}{\pi^2}\left(1 + \frac{4\lambda}{d}\right). \qquad (4.17) $$

The dependence of τ on the sample thickness is thus parabolic, with a linear correction proportional to the extrapolation length.

In weak anchoring regime, the amplitude of the orientational director fluctuations at the surface is substantial, so that in addition to the bulk elastic, also surface elastic torque acts on the director and determines the relaxation dynamics. In this case the extrapolation length is larger or comparable to sample thickness and using (4.15), the eigenvalues q_{zN} can be obtained. They are reduced in comparison to strong anchoring and the relaxation time of the fundamental fluctuation mode is [8]

$$\tau_0 \approx \frac{\eta}{K} \frac{\lambda}{2} d. \tag{4.18}$$

As one can see from (4.17) and (4.18), for both weak and strong anchoring regimes the relaxation rates of the fundamental orientational fluctuation modes depend on the sample thickness and the interaction with the aligning substrate. This means that by measuring the relaxation times as a function of the sample thickness, surface anchoring coefficients can be obtained. This idea has already been successfully implemented for several different systems [8, 26, 27]. The great advantage of this method is that no external torques act on the liquid crystal during the measurements, which could influence the liquid crystal and the results.

So far, only the dynamics of the fundamental director fluctuation mode has been considered. This is indeed the major contribution to scattering if the sample is very thin. However, often also higher modes contribute significantly to the relaxation dynamics in the sample. The relaxation rates for all modes are determined by the secular equations (4.15), whereas the amplitudes still need to be calculated. Using (4.10), the intensity I_N of the N-th mode, which is proportional to the square of the scattered electric field, can be written for a planar sample as [28]:

$$I_N \propto \left(\frac{\sin(\frac{q_{zN}-q}{2}d)}{(\frac{q_{zN}-q}{2}d)} \right)^2 \frac{d^2}{q_{zN}^2}. \tag{4.19}$$

From (4.19) it is easy to see that if the thickness of the sample d is much larger than the wavelength of the scattered light, the amplitude becomes a Dirac delta function and only the fluctuation mode with $q = q_z$ is observed. Another case where just one fluctuation mode is observed is for very small sample thickness, where the fundamental mode prevails. In the intermediate regime, however, where the thickness of the sample is comparable to the wavelength of light, several modes contribute to scattering of light, each with the corresponding amplitude. Intensities of the scattered light on the first three fluctuation relaxation modes as a function of sample thickness are shown in Fig. 4.3. The

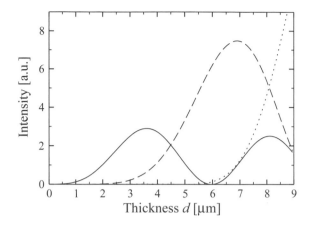

Fig. 4.3 Intensities of the scattered light on the first three orientational fluctuation modes as a function of sample thickness (the *solid line* $N = 0$, the *dashed line* $N = 1$, and the *dotted line* $N = 2$). The fundamental mode prevails for small sample thickness (up to ≈ 3 μm), where the intensity of the next mode begins to increase. The scattering vector in the calculation was $q = 1.5$ μm^{-1} and the extrapolation length $\lambda = 450$ nm

autocorrelation function that one measures in this case is a composite of several exponentially decaying functions and there exists a whole distribution of relaxation times, which are of the same order of magnitude. Therefore the autocorrelation function can be reasonably well fitted with a stretched exponential decay function

$$\left(g^2(\mathbf{q}, t) - 1\right) \propto \exp(-(t/\tau_r)^s), \tag{4.20}$$

where τ_r is an average relaxation time and the stretching exponent s differs from 1.

How the different fluctuation modes contribute to the relaxation time can also be numerically calculated and the effective relaxation time obtained as a sum of several contributing fluctuation modes is shown in Fig. 4.4. The relaxation time first increases accordingly to the anchoring regime (linear in the case of weak and parabolic in the case of strong anchoring) until the higher modes with a smaller relaxation time prevail and the net relaxation time is reduced. With increasing sample thickness it approaches the bulk value.

Such behaviour has also been experimentally observed and will be shown in the following section. Further examples where the stretched exponential function needs to be used is either in disoriented nematic samples with an orientational distribution of the director ($s \approx 0.8$), or in liquid crystal droplets with a distribution of scattering vectors ($s \approx 0.2$).

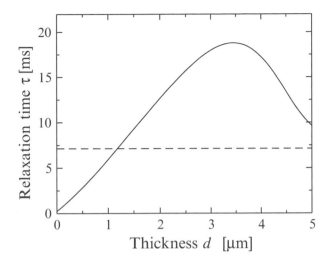

Fig. 4.4 Calculated relaxation time τ as a function of sample thickness if the first three fluctuation modes are taken into account. The extrapolation length was $\lambda = 450\,\text{nm}$, the scattering vector $q = 1.5\,\mu\text{m}^{-1}$ and $q_x = 0.6\,\mu\text{m}^{-1}$. The *dashed line* indicates the bulk relaxation time. Reprinted with permission from [8]. Copyright by the American Physical Society

4.6.2 Relaxation Dynamics in PDLCs

In spherical geometry the situation is more complex as in planar cells since the equilibrium configuration of the director is not uniform. In radial spherical droplets there is a singularity in the centre of the droplet and therefore the lowest eigenvalue of the fluctuations is of the order of $1/R$ [24]. The bulk elasticity thus always dominates the relaxation rate, regardless of the anchoring strength.

In the case of bipolar structure (Fig. 4.1), a new low frequency mode appears in the spherical droplets, in which reorientation of the average droplet direction occurs. As there is no restoring torque for this type of motion, this mode represents free diffusion of the droplet orientation. This can, however, be strongly influenced by external electric fields.

4.6.3 Relaxation Dynamics in HPDLCs

Because of the symmetry of the H-PDLC formation process, liquid crystal domains phase separated from the polymer host are axially symmetric ellipsoids, which are usually compressed in the direction of the symmetry axis. It is reasonable to expect that for such oblate ellipsoids the spectrum of director fluctuations is intermediate between the spherical and the planar geometry and to facilitate the analysis, we examine the situation of finite cylindrical cavities of the radius R and length $d < R$.

In most of the H-PDLC materials a preferential orientation of the nematic director field $\mathbf{n}(\mathbf{r})$ is pointing along the symmetry axis $\mathbf{e}_z$ of the droplets (see Fig. 4.2). The corresponding structure can be described by hybrid boundary conditions: i.e. the anchoring on base planes of the cylindrical cavities is assumed to be homeotropic, while the anchoring on side walls is taken to be planar. Consequently, the equilibrium director configuration is spatially homogeneous and is given by $\mathbf{n}_0 = (0, 0, 1)$. The orientational fluctuations take place in the xy plane and have a form of Bessel functions of the first kind in the radial direction and a form of sinusoidal waves along the z axis. For strong but finite surface anchoring characterized by $\lambda \ll d$, the relaxation rate of the fundamental eigenmode is given by

$$\frac{1}{\tau_0} = \frac{K}{\eta}\left(\left(1 - \frac{2\lambda}{R}\right)\left(\frac{2.4}{R}\right)^2 + \left(1 - \frac{4\lambda}{d}\right)\left(\frac{\pi}{d}\right)^2\right). \qquad (4.21)$$

Application of external electric field in the direction parallel to the liquid crystal rich planes breaks the intrinsic axial symmetry of the nematic director field within the droplets. This causes a slowing down of orientational fluctuations similar to the case of Freedericksz transition in planar nematic cells [29–31]. For $\mathbf{E} = E_0\mathbf{e}_x$ the dynamic equation for the fluctuations (4.5) becomes

$$K\nabla^2\delta\mathbf{n}(\mathbf{r},t) + \varepsilon_0\varepsilon_a E_0^2(\mathbf{e}_x\delta\mathbf{n})\mathbf{e}_x = \eta\frac{\partial\delta\mathbf{n}}{\partial t}, \qquad (4.22)$$

and the relaxation rate of fundamental director fluctuations parallel to the $\mathbf{e}_x$ axis is given by

$$\frac{1}{\tau_{0,x}}(E_0) = \frac{1}{\tau_0} - \frac{\varepsilon_0\varepsilon_a E_0^2}{\eta} = \frac{\varepsilon_0\varepsilon_a}{\eta}(E_c^2 - E_0^2). \qquad (4.23)$$

Dielectric interaction destabilizes the homogeneous director configuration $\mathbf{n}_0(\mathbf{r})$ and at $E_c = (\eta/\tau_0\varepsilon_0\varepsilon_a)^{1/2}$ the fluctuations along $\mathbf{e}_x$ axis "freeze" and correspondingly a second order structural transition takes place. Contrary to this, the relaxation rate of fluctuations $\delta\mathbf{n}$ parallel to the $\mathbf{e}_y$ axis is not influenced by $\mathbf{E} = E_0\mathbf{e}_x$ and remains the same as in the absence of the field (4.21).

Above the structural transition, the equilibrium director profile $\mathbf{n}_0(\mathbf{r})$ is no longer orthogonal to the field. This feature modifies its dielectric coupling to the field and as a result for $E_0 > E_c$ the effect of the field is that it stabilizes the structure by suppressing the fluctuations. As in (4.23), in the vicinity of E_c, the response of the critical mode is expected to be proportional to $E_0^2 - E_c^2$.

4.7   Dynamic Light Scattering Experiments

The orientational director fluctuation modes have been described and discussed for thin planar cells and liquid crystal droplets. Now experimental observations for these two special geometries will be presented and the implementation of the experiment for measurements of anchoring energy and investigations of the structural transitions of the nematic director field will be given.

4.7.1   Planar Samples

It has been shown in previous sections that studying the orientational director dynamics yields information about the interaction of a liquid crystal with the aligning surface. Therefore dynamic light scattering experiments on planar thin samples can be used to determine the anchoring coefficients of a liquid crystal on a given substrate. Measuring the relaxation times in order to obtain the anchoring strengths has some major advantages in contrast to conventional methods. First of all, the director is in an undistorted state as no external fields act on the liquid crystal during the measurements. Moreover, the same sample can be used for measurements of both the azimuthal and zenithal anchoring energy coefficients, which makes the comparison between the two obtained coefficients more reliable.

To obtain the anchoring strengths, the relaxation time of the fundamental orientational fluctuation mode needs to be measured as a function of sample thickness. This can best be done by using a wedge-like cell. The gradient of the thickness should be small, so that no additional elastic effects appear. Also the overall sample thickness needs to be small, especially for the measurements of strong anchoring, as it needs to be comparable to the extrapolation length. Typical thickness range is thus $d \approx 1 - 3\,\mu\mathrm{m}$ for weak anchoring, and $d \approx 0.2 - 2\,\mu\mathrm{m}$ for strong anchoring. The thickness of the sample needs to be very accurately measured and this is done using a spectrophotometer. Observing the transmission of an empty cell as a function of the wavelength of the incident light, the positions of transmission peaks and dips yields the sample thickness with a sub-wavelength resolution. In a planar wedge-like cell, the director needs to be oriented parallel to the thickness gradient, which is achieved by using a proper aligning layer. Depending on the preparation of the substrate, both strong or weak anchoring regimes can be studied.

The dynamic light scattering experiment is performed using a standard photon-correlation setup. The light source is a laser with a good temporal stability – in the experiments described below, the laser was a He–Ne laser operating at a wavelength of 632.8 nm. The intensity correlation function $g^{(2)}$ (4.11) is measured using a correlator (ALV-5000) that enables measurements on a large timescale (in our case from $10^{-8} - 10^3\,\mathrm{s}$). As shown in previous sections, in the heterodyne regime of the measurement, the decay rate of the obtained autocorrelation function equals the relaxation rate of the chosen fluctuation mode. The relaxation rate is then measured as a function of sample thickness by moving the sample. A heating stage can be used, either to stabilise the temperature, or to observe temperature dependence of fluctuation modes and surface anchoring.

For measurements of the two anchoring coefficients the same sample can be used in different scattering geometries. For azimuthal anchoring measurements, the twist-bend fluctuation mode is observed and this is achieved by orienting the director $\mathbf{n}$ perpendicular to the scattering plane. The polarisation of the incident light is parallel to the director and the polarisation of the detected beam is chosen to be in the scattering plane. Small scattering angle reduces the contribution of other fluctuation modes to the relaxation time, so that only the fundamental twist mode is detected.

Zenithal anchoring coefficient can be determined if the director and both the polarisations are in the scattering plane. In this case the contribution of the statically scattered light on the impurities and defects in the sample as well as on the sample boundaries is significant. Also, a contribution of the wave vector component parallel to the sample boundaries is present. However, this term is thickness independent, meaning that it does not considerably disturb the determination of the anchoring coefficient. A detailed description and discussion of the two scattering geometries can be found in [8].

A typical measured autocorrelation function of the dynamically scattered light is shown in Fig. 4.5. It was obtained by using a nematic liquid crystal 4-n-pentyl-4′-cyanobiphenyl (5CB) and rubbed Nylon as aligning layer at a cell thickness of ≈2 µm [8]. The dots show the measured curve and the solid line the stretched exponential fit (4.20). The amplitude of the $g^{(2)}$ shows that the measurement was performed in a heterodyne regime as the intensity of the light scattered by fluctuations was fairly small. Therefore the characteristic

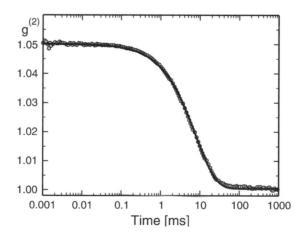

Fig. 4.5 Measured autocorrelation function of dynamically scattered light by orientational fluctuations in nematic liquid crystal 5CB. The *circles* represent the measured data and the *solid line* the stretched exponential fit to the data with the relaxation time $\tau = 7.86 \pm 0.05$ s and stretching exponent $s = 0.84 \pm 0.03$. Reprinted with permission from [8]. Copyright by the American Physical Society

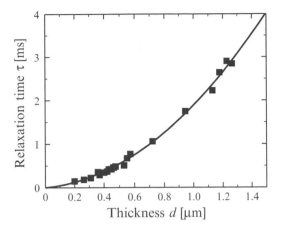

Fig. 4.6 Orientational fluctuation relaxation time as a function of sample thickness for nematic liquid crystal 5CB on rubbed Nylon with strong anchoring at the boundaries. The *squares* are the measured data and the *solid line* best fit of the parabolic approximation. Reprinted with permission from [8]. Copyright by the American Physical Society

decay time of the autocorrelation function equals the fluctuation relaxation time. Stretching exponent $s < 1$ indicates that more than just one fluctuation mode was observed. The relaxation times were then measured as a function of sample thickness and depending on the preparation of the aligning surface, both weak and strong anchoring as well as intermediate regime were studied.

In the case of strong anchoring, parabolic dependence of the relaxation time on the sample thickness is observed and is shown in Fig. 4.6. By fitting (4.17) to the measured data, two parameters can be determined independently: the first one is the combination of the viscoelastic constants η/K, and the second is the extrapolation length λ. The first parameter can be compared to the values obtained by other authors and an agreement implies that the method and the obtained extrapolation length are reliable. Extrapolation length in this case was found to be $\lambda = 50 \pm 10$ nm, which corresponds to the anchoring energy of $W_\varphi = (5.6 \pm 1.2) \times 10^{-5}$ J m^{-2}. The limitation of the measurement is the thickness of the cell as only extrapolation lengths comparable to the sample thickness can be measured – meaning a few tens of nanometres.

On the other hand, in weak anchoring regime the extrapolation lengths can be measured as long as stability of the alignment is assured. For such samples, linear dependence as predicted by (4.18) is observed and an example of obtained curves is shown in Fig. 4.7. The aligning Nylon layer was only weakly rubbed and the liquid crystal was the same as in the previous case (5CB). By fitting (4.18) to the obtained data, the extrapolation length can be obtained if the ratio K/η is known, and for the example shown in Fig. 4.7, $\lambda = 450 \pm 80$ nm, corresponding to the anchoring energy of $W_\varphi = (6.2 \pm 1.2) \times 10^{-6}$ J m^{-2}.

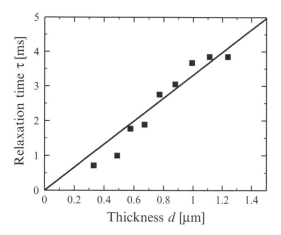

Fig. 4.7   Orientational fluctuation relaxation time as a function of sample thickness for nematic liquid crystal 5CB on rubbed Nylon with weak anchoring at the boundaries. The *squares* are the measured data and the *solid line* best fit of the linear function

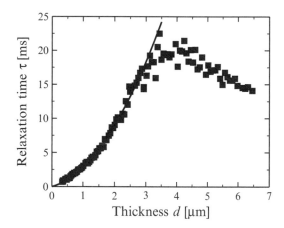

Fig. 4.8   Orientational fluctuation relaxation time as a function of sample thickness for nematic liquid crystal 5CB on rubbed Nylon. In the thinner part of the cell, parabolic behaviour is observed, whereas in the thicker part of the sample, higher fluctuation modes prevail and the relaxation time is reduced. It slowly approaches the bulk value. Reprinted with permission from [8]. Copyright by the American Physical Society

So far, only the fundamental fluctuation mode has been discussed and used for determining the anchoring coefficient. In this case, either linear or parabolic dependence of the relaxation time on the sample thickness is observed, depending on the anchoring strength. However, for larger sample thicknesses, above a few μm, a significant deviation from the relaxation time of the fundamental mode is observed (Fig. 4.8). The data shown were obtained using 5CB as liquid

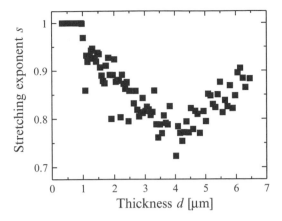

Fig. 4.9 The stretching exponent s as a function of sample thickness for nematic liquid crystal 5CB on rubbed Nylon. In the thinner part of the cell, $s \approx 1$ as only the fundamental mode is observed. The value decreases with increasing influence of the second fluctuating mode and when the latter prevails, s increases again. Reprinted with permission from [8]. Copyright by the American Physical Society

crystal and rubbed Nylon as aligning layer. As one can observe, the measured dependence resembles the theoretically predicted one (Fig. 4.4).

Also the stretching exponent can be obtained from the data and the results are shown in Fig. 4.9. They can be well explained by the mixing of different fluctuation modes and their relative contributions to the scattered light for small sample thickness. When only the fundamental mode is observed, $s = 1$ and in this region the anchoring coefficients can be determined fairly straightforward. The decrease in s is due to the contribution of the second fluctuating mode and when the second mode prevails, the s increases again.

Besides the azimuthal anchoring coefficients W_φ, also the zenithal coefficients W_ϑ can be measured. The principle of obtaining the values is the same, although the measurements are much more demanding due to a large amount of statically scattered light. Since the two coefficients can be determined using the same sample, a reliable comparison between the two coefficients can be made and in contrast to some previously reported results [32], the anchoring coefficients were found to differ only by a factor of 2 or less. Using a heating stage, the values of the two coefficients can be acquired in the whole regime of the nematic phase as shown in Fig. 4.10. The ratio of the constants was found to be almost temperature independent [27]. In fact, when comparing the temperature dependence of the ratio of the two anchoring coefficient with the ratio of the corresponding Frank elastic constants taken from the literature [33], surprisingly good agreement is found (Fig. 4.11). This observation indicates that the origin of the anchoring anisotropy could be found in the anisotropy of

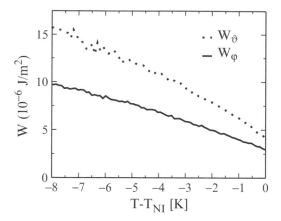

Fig. 4.10 Anchoring energy coefficients W_φ (*solid line*) and W_ϑ (*dotted line*) as a function of temperature for nematic 5CB on rubbed Nylon. T_{NI} is the transition temperature from the nematic to isotropic phase. Reprinted with permission from [27]. Copyright by the American Physical Society

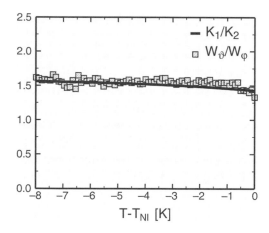

Fig. 4.11 A comparison of the ratios of the two anchoring strengths W_ϑ/W_φ (*squares*) with the ratio of the corresponding Frank elastic constants K_1/K_2 (*solid line*). Reprinted with permission from [27]. Copyright by the American Physical Society.

the internal interactions in a thin surface layer of the liquid crystal rather than in the anisotropy of the interactions with the aligning surface.

4.7.2   HPDLCs

A similar setup was also used to study the orientational fluctuations in liquid crystal domains formed in 5-μm-thick H-PDLC transmission gratings. The pitch of the gratings was $\Lambda = 0.78$ μm for the samples polymerized using visible (VIS) and 1 μm for the samples polymerized using the UV laser irradiation [9].

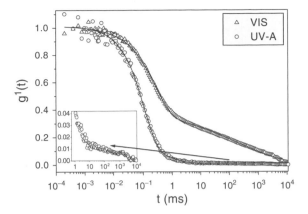

Fig. 4.12 First order autocorrelation functions $g^{(1)}(t')$ of scattered light detected in VIS and in UV samples at small scattering angles at room temperature. *Solid lines* are fits to stretched double exponential decay. Reprinted with permission from [9]. Copyright by the American Physical Society

Polarisation of the incident laser beam was parallel and polarisation of the scattered light was perpendicular to the scattering plane. For each scattering angle, the sample was rotated so that the scattering wave vector $\mathbf{q}_s$ was always parallel to the glass substrates and was either parallel to the grating wave vector $\mathbf{K}_g = (2\pi/\Lambda)\mathbf{e}_z$, or perpendicular to it. Because of the presence of strong elastic scattering acting as a local oscillator at all scattering angles, the measurements were performed in the heterodyne detection regime.

Figure 4.12 shows two typical autocorrelation functions $g^{(1)}(t')$ measured at room temperature. For both types of the mixtures, two dynamic modes are observed: a fast nearly exponential mode with a relaxation time τ_f in the range of 10^{-3} s, and a slow remarkably non-exponential mode with τ_s in the range of $10 - 10^3$ s and stretching exponent $s \approx 0.2 - 0.5$. In the VIS samples, the amplitudes of both modes are nearly identical, while in the UV samples, the slow mode is much less pronounced (see inset). The fast mode is related to intradroplet or intrapore orientational fluctuations, while the slow mode is attributed to the orientational diffusion of the preferential director orientation due to the randomness and irregularities of the polymer–LC interface [34–37]. The low amplitude of the slow mode in the UV samples suggests that in these mixtures the nematic director field has a well-defined preferential configuration, while in the VIS mixtures several configurations can be realized with similar probabilities.

The magnitude of $\mathbf{q}_s$ at which $1/\tau_f$ starts to increase by increasing $\mathbf{q}_s$ depends on the droplet size and is given by $q_{\min} \sim \pi/L$, where L is the characteristic dimension of confinement in the direction of $\mathbf{q}_s$ (see 4.19). As evident from Fig. 4.13, for $\mathbf{q}_s \parallel \mathbf{K}_g$ this happens at significantly larger value of $\mathbf{q}_s$ than for $\mathbf{q}_s \perp \mathbf{K}_g$, which indicates that the droplets are squeezed in the direction of $\mathbf{K}_g$

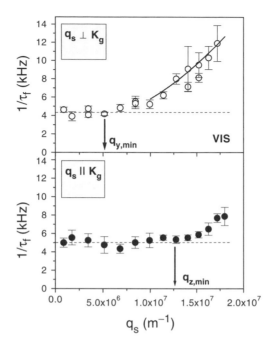

Fig. 4.13 Dispersion relation of the fast relaxation rate measured in VIS sample at room temperature. *Solid line* is a fit to (4.24). *Vertical arrows* indicate starting points of the increasing relaxation rate according to the size of the droplets as deduced from the SEM study. The *dashed lines* are guide to the eye. Reprinted with permission from [9]. Copyright by the American Physical Society

(Fig. 4.2). This is in agreement with the structure of these samples as revealed from our SEM study [38, 39]. According to the SEM images, the droplets are oblate ellipsoids of about $250 \times 600 \times 600$ nm in size. One can notice that the values of $q_{y,\min}$ and $q_{z,\min}$ corresponding to this size well match the starting points of increasing relaxation rate $1/\tau_f$. This result demonstrates that dynamic light scattering is a convenient alternative method to probe the size and the shape of liquid crystal domains in H-PDLCs. For $\mathbf{q}_s \ll q_{\min}$, the scattering signal is associated with the fundamental eigenmode of the fluctuations, the relaxation rate of which is approximately given by (4.21). In another limit, for $\mathbf{q}_s \gg q_{\min}$, again only one eigenmode is expected to dominate the scattering signal and a bulk-like parabolic dispersion given by the expression

$$\frac{1}{\tau(\mathbf{q}_s)} = \frac{K}{\eta}\left(\mathbf{q}_s^2 + \left(1 - \frac{4\lambda}{d}\right)\left(\frac{\pi}{d}\right)^2\right), \qquad (4.24)$$

should be observed in $\mathbf{q}_s \perp \mathbf{K}_g$ geometry. The best fit of experimental data obtained for $\mathbf{q}_s > 2q_{y,\min}$ to (4.24) is shown as a solid line in Fig. 4.13.

From the analysis described above, one can deduce the viscoelastic and surface anchoring parameters of the liquid crystalline material phase separated from the polymer matrix, which in our VIS samples have the values $K/\eta = (0.3 \pm 0.06) \cdot 10^{-10} \, \text{m}^2 \, \text{s}^{-1}$ and $\lambda = 65 \pm 35 \, \text{nm}$. The observed diffusivity K/η is almost four times lower than the diffusivity of a pure liquid crystal used to prepare the mixture. This result demonstrates that to properly explain the dynamic properties of the H-PDLC gratings, such as switching-on and -off times, it may not be relevant to take viscoelastic parameters of the liquid crystal substances as used for the mixture, but one needs to measure the properties of a nematogenic material formed in the phase separation process. Since the scattering experiments require no special sample treatment, it is one of the most convenient techniques to perform this task. The observed decrease of K/η is attributed to the increased viscosity of a liquid crystal medium due to the presence of dissolved parts of the polymer chains [26,40]. The observed value of λ results in $W = (2.1 \pm 1.1) \cdot 10^{-4} \, \text{J} \, \text{m}^{-2}$, which corresponds to a quite strong anchoring.

Figure 4.14 shows the dependence of the diffraction efficiency η_d and relaxation time τ_f on the magnitude of the external electric field applied along the

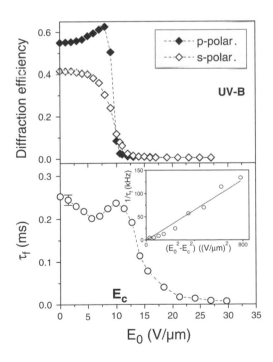

Fig. 4.14 Diffraction efficiencies and relaxation time of the fast scattering mode as functions of amplitude of external electric field measured in the UV grating at room temperature. *Solid line* in the inset is a fit to linear dependence. The *dashed lines* are a guide for the eye. Reprinted with permission from [9]. Copyright by the American Physical Society

$\mathbf{e_x}$ axis (see Fig. 4.2). The values of τ_f were obtained in the $\mathbf{q_s} \perp \mathbf{K_g}$ scattering geometry at $\theta_s = 60°$, so that contribution from the fundamental mode prevailed the signal. The threshold field E_c of a H-PDLC is related to the region of a steep decrease of diffraction efficiency with increasing field amplitude. Analyzing the director fluctuations, one can observe that below E_c, a nonmonotonous behaviour of $\tau_f(E_0)$ takes place, while above E_c, the relaxation time decreases with increasing E_0. In the graph for the diffraction efficiency $\eta_d(E_0)$, one can notice that just above E_c there is an interval in which the diffraction efficiency for s-polarisation is significantly larger than for p polarisation. This is associated with the intermediate state in which the droplet directors are already preferentially lying in the xy plane, but are not yet fully aligned along the electric field. For that reason the refractive index mismatch for s polarisation is larger than for p polarisation. At even higher fields, the structure is close to $\mathbf{n(r)} = (1, 0, 0)$ and the diffraction efficiency for both polarisations is about the same. All these observations suggest that a Freedericksz-like structural transition from $\mathbf{n(r)} \sim (0, 0, 1)$ to $\mathbf{n(r)} \sim (1, 0, 0)$ takes place at $E = E_c$.

According to (4.23), in the vicinity of E_c the response of the critical mode is expected to be governed by $E_0^2 - E_c^2$. The best fit of the data obtained for $E_0 > E_c$ to the relation $1/\tau_f \propto (E_0^2 - E_c^2)$ is shown as a solid line in the inset of Fig. 4.14 and a good agreement is found. On the contrary, for $E_0 < E_c$, the experimentally observed dependence of $\tau_f(E_0)$ is quite different from the theoretically expected behaviour. The potential reasons for this are numerous. At selected scattering angle $\theta_s = 60°$ our scattering experiment probes both the noncritical $\delta\mathbf{n}\|\mathbf{e_y}$ and the critical fluctuation $\delta\mathbf{n}\|\mathbf{e_x}$ mode, so that the obtained values of τ_f correspond to an average relaxation time determined by relative amplitudes of these two modes. The size of the droplets is not uniform, so that the value of E_c varies from droplet to droplet and a convolution of several critical processes is monitored. The critical phenomena are consequently smoothed. In addition, for a significant part of the droplets the average droplet director is tilted from the $(0, 0, 1)$ direction, so that they exhibit a noncritical fluctuation dynamics associated to $1/\tau_0 \propto E_0^2$. These 'misaligned' droplets are in our opinion the main reason why at low fields τ_f decreases by increasing field and a critical slowing down becomes pronounced only when the scattering from 'aligned' droplets start to prevail in the dynamic light scattering response.

By changing the temperature of H-PDLC gratings, an interesting dependence of $\eta_d(T)$ is found, exhibiting notable similarity to the behaviour of $\eta_d(E_0)$, shown in Fig. 4.14. This is illustrated in Fig. 4.15, which shows the temperature dependencies of η_d and τ_f for a UV grating. Above some specific temperature the diffraction efficiency for s polarisation starts to exceed the diffraction efficiency for p polarisation.

This situation is observed up to the nematic-isotropic (N-I) phase transition at which η_s and η_p become nearly the same. The corresponding values of τ_f

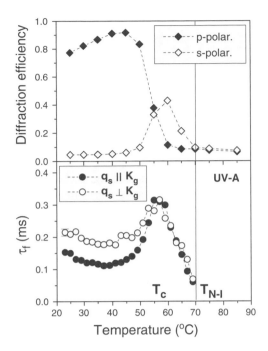

Fig. 4.15 Temperature dependence of diffraction efficiencies and relaxation time of the fast scattering mode detected in the UV grating in the absence of external field. *Vertical line* denotes the nematic to isotropic phase transition. The *dashed lines* are a guide for the eye. Reprinted with permission from [9]. Copyright by the American Physical Society

obtained at $\theta_s = 40°$ show that the cross-over is associated with the slowing down of the fundamental fluctuation mode, which is detected in the vicinity of some critical temperature T_c. Above T_c, the relaxation time τ_f decreases with increasing temperature.

An additional feature related to T_c is that a strongly-non-exponential dynamic mode, which is observed in $g^{(1)}(t)$ at room temperature (Fig. 4.12), vanishes above T_c. This means that below T_c several configurations of nematic director field can appear with similar probabilities, while above T_c, there exist a well-defined preferential configuration. This is possible only if the role of polymer–LC interface in fluctuation dynamics is strongly modified at T_c. A strong modification of interface effects can be attributed to a gradual melting of the nematic phase from the surface of the droplets. The polymer–LC interface formed during photopolymerisation is usually quite rough [41–43] and, as found from the measurements of the viscoelastic parameters, dangling parts of polymer chains are very probably present in the LC phase. This results in a local decrease of the N-I transition temperature at the interface. For PDLC droplets with a diameter of around 10 μm, such a surface-induced melting was observed by optical microscopy by Admundson et al., who found that melting induces structural transition that dramatically affects the electrooptic response of the medium [44].

The reason for reorientation of the droplet director by surface melting in our samples is that the polymer surface prefers a homeotropic alignment of the oblate droplets, while isotropic-nematic liquid crystal interface is known to prefer a tilted, nearly planar director anchoring [32, 45]. Consequently at some critical temperature T_c, at which a more or less closed shell of isotropic LC phase is formed between the droplet surface and the central region of the nematic phase, the structural transition of director field from $\mathbf{n}(\mathbf{r}) \sim (0, 0, 1)$ to $\mathbf{n}(\mathbf{r}) \sim (\cos \varphi, \sin \varphi, 0)$ takes place [46]. The value of φ, which is associated with the axis of resulting bipolar configuration, is determined by the shape and surface topography of a particular droplet. The reorientation causes an increase of refractive index mismatch for s and a decrease of refractive index mismatch for p polarisation, so the corresponding diffraction efficiencies are strongly modified with respect to their values at $T < T_c$.

As in the case of electric field induced structural transition, the temperature induced structural transition is also associated with critical slowing down of the director fluctuations. However, as reorientation of the droplet director can occur along any direction in the xy plane, both $\delta \mathbf{n} \| \mathbf{e}_y$ and $\delta \mathbf{n} \| \mathbf{e}_x$ fundamental modes exhibit a critical behaviour of $1/\tau_0$ in the vicinity of T_c and consequently a slowing down detected in the dynamic light scattering experiments is more pronounced than in case of electric field induced transition.

4.8   Conclusions

We have shown how the confinement affects the orientational fluctuations in liquid crystals and how measurements of their relaxation rates by dynamic light scattering can be used to obtain the information on the interaction of liquid crystal with confining surfaces. The analysis of the fluctuation relaxation times as a function of sample thickness in thin wedge-shaped cells provides one of the most accurate ways to determine the azimuthal and zenithal anchoring energy coefficients of liquid crystals in contact with various substrates. Our measurements also demonstrate that dynamic light scattering is a very powerful tool for studying the structural and dynamic properties of PDLCs, H-PDLCs and related confined liquid crystal media. It allows one to determine the size and shape of liquid crystalline droplets and to measure viscoelastic and surface anchoring parameters of the liquid crystal material phase separated from the polymer matrix. This information is prerequisite to optimize the electrooptic performance of the material. The technique can in principle be used in-situ, during the photopolymerisation reaction, which provides a way to control the phase separation kinetics, structural morphology and associated dynamic properties in the course of composite formation process.

While the properties of the fast dynamic mode related to intrapore orientational fluctuations of the director field are quite well understood, the origin of various slow dynamic processes is for the moment still far from being resolved. The observed characteristics of the slow mode are typically very specific for a selected system and strongly depend on the morphology and composition of the confining medium. The slow process usually exhibits a profound non-exponetial relaxation, which is attributed to the structural randomness of the material. Further studies in confined liquid crystal media with well-resolved morphological details are needed to elucidate the nature of different slow modes.

References

[1] de Gennes, P.G. and Prost, J. (1993). *The Physics of Liquid Crystals.* Clarendon, Oxford.

[2] de Jeu, W.H. (1980). *Physical Properties of Liquid Crystalline Materials.* Gordon and Breach, London.

[3] Orsay Liquid Crystal Group (1969). Quasielastic Rayleigh scattering in nematic liquid crystals, *Phys. Rev. Lett.* 22:1361–1363.

[4] Orsay Liquid Crystal Group (1969). Dynamics of fluctuations in nematic liquid crystals, *J. Chem. Phys.* 51:816–822.

[5] Crawford, G.P. and Žumer, S., editors (1996). *Liquid Crystals in Complex Geometries.* Taylor and Francis, London.

[6] Mertelj, A. and Čopič, M. (1998). Surface-dominated orientational dynamics and surface viscosity in confined liquid crystals, *Phys. Rev. Lett.* 81:5844–5847.

[7] Mertelj, A. and Čopič, M. (2000). Dynamic light scattering as a probe of orientational dynamics in confined liquid crystals, *Phys. Rev. E* 61: 1622–1628.

[8] Vilfan, M., Mertelj, A., and Čopič, M. (2002). Dynamic light scattering measurements of azimuthal and zenithal anchoring of nematic liquid crystals, *Phys. Rev. E* 65:041712(1–7).

[9] Drevenšek-Olenik, I., Jazbinšek, M., Sousa, M.E., Fontecchio, A.K., Crawford, G.P., and Čopič, M. (2004). Structural transitions in holographic polymer-dispersed liquid crystals, *Phys. Rev. E* 69:051703(1–9).

[10] Rapini, A. and Papoular, M. (1969). Distorsion d'une lamelle nématique sous champ magnétique conditions d'ancrage aux parois, *J. Phys.* 30 C4:54–56.

[11] Berne, B.J. and Pecora, R. (2000). *Dynamic Light Scattering.* Dover, New York.

[12] Doane, J.W., Vaz, N.S., Wu, B.G., and Žumer, S. (1986). Field controlled light scattering from nematic microdroplets, *Appl. Phys. Lett.* 48:269–271.

[13] Dubois-Violette, E. and Parodi, O. (1969). Emulsions nématiques effects de champ magnétiques et effects piézoélectriques, *J. Phys.* 30 C4:57–64.

[14] For a review see for instance Bunning, T.J., Natarjan, L.V., Tondiglia, V.P., and Sutherland, R.L. (2000). Holographic polymer-dispersed liquid crystals (H-PDLCs), *Annu. Rev. Mater. Sci.* 30:83–115.

[15] Crawford, G.P. (2003). Electrically switchable Bragg gratings, *Opt. Photon. News* 14:54–59.

[16] Bowley, C.C. and Crawford, G.P. (2000). Diffusion kinetics of formation of holographic polymer-dispersed liquid crystal display materials, *Appl. Phys. Lett.* 76:2235–2237.

[17] Kyu, T., Nwabunma, D., and Chiu, H.-W. (2001). Theoretical simulation of holographic polymer-dispersed liquid-crystal films via pattern photopolymerization-induced phase separation, *Phys. Rev. E* 63:061802(1–8).

[18] Sutherland, R.L., Tondiglia, V.P., Natarajan, L.V., Bunning, T.J., and Adams, W.W. (1994). Electrically switchable volume gratings in polymer-dispersed liquid-crystals, *Appl. Phys. Lett.* 64:1074–1076.

[19] Sutherland, R.L., Tondiglia, V.P., Natarajan, L.V., and Bunning, T.J. (2001). Evolution of anisotropic reflection gratings formed in holographic polymer-dispersed liquid crystals, *Appl. Phys. Lett.* 79:1420–1422.

[20] Sutherland, R.L. (2002). Polarization and switching properties of holographic polymer-dispersed liquid-crystal gratings. I. Theoretical model, *J. Opt. Soc. Am. B* 19:2995–3003.

[21] Vardanyan, K.K., Qi, J., Eakin, J.N., De Sarkar, M., and Crawford, G.P. (2002). Polymer scaffolding model for holographic polymer-dispersed liquid crystals, *Appl. Phys. Lett.* 81:4736–4738.

[22] Stallinga, S., Wittebrood, M.M., Luijendijk, D.H., and Rasing, Th. (1996). Theory of light scattering by thin nematic liquid crystal films, *Phys. Rev. E* 53:6085–6092.

[23] Ziherl, P. and Žumer, S. (1996). Nematic order director fluctuations in cylindrical capillaries, *Phys. Rev. E* 54:1592–1598.

[24] Kelly, J.R. and Palffy-Muhoray, P. (1997). Normal modes of a radial nematic droplet, *Phys. Rev. E* 55:4378–4381.

[25] Wittebrood, M.M., Rasing, Th., Stallinga, S., and Muševič, I. (1998). Confinement effects on the collective excitations in thin nematic films, *Phys. Rev. Lett.* 80:1232–1235.

[26] Vilfan, M., Drevenšek Olenik, I., Mertelj, A., and Čopič, M. (2001). Aging of surface anchoring and surface viscosity of a nematic liquid crystal on photoaligning poly(vinyl-cinnamate), *Phys. Rev. E* 63:061709(1–5).

[27] Vilfan, M. and Čopič, M. (2003). Azimuthal and zenithal anchoring of nematic liquid crystals, *Phys. Rev. E* 68:031704-1–031704-5.

[28] Mertelj, A. and Čopič, M. (1998). Dynamic light scattering in nematic liquid crystals in confined geometries, *Mol. Cryst. Liq. Cryst.* 320: 287–299.

[29] Eidner, K., Lewis, M., Vithana, H.K.M., and Johnson, D.L. (1989). Nematic-liquid-crystal light-scattering in a symmetry-breaking external-field, *Phys. Rev. A* 40:6388–6394.

[30] Galatola, P. and Rajteri, M. (1994). Critical-noise measurement near Freedericksz transitions in nematic liquid-crystals, *Phys. Rev. E* 49: 623–628.

[31] Drevenšek-Olenik, I., Jazbinšek, M., and Čopič, M. (1999). Localized soft mode at optical-field-induced Freedericksz transition in a nematic liquid crystal, *Phys. Rev. Lett.* 82:2103–2106.

[32] Jérôme, B. (1998). Surface Alignment. In Demus, D., Goodby, J., Gray, G.W., Spiess, H.-W., and Vill, V., editors, *Handbook of Liquid Crystals*, Vol. 1, p. 535. Wiley, New York; and references within.

[33] Chen, G.-P., Takezoe, H., and Fukuda, A. (1989). Determination of $K_i(i = 1 - 3)$ and $\mu_j(j = 2 - 6)$ in 5CB by observing the angular dependence of Rayleigh line spectral widths, *Liq. Cryst.* 5:341–347.

[34] Bellini, T., Clark, N.A., and Schaefer, D.W. (1995). Dynamic light-scattering study of nematic and smectic-A liquid crystal ordering in silica aerogel, *Phys. Rev. Lett.* 74:2740–2743.

[35] Mertelj, A. and Čopič, M. (1997). Evidence of dynamic long-range correlations in a nematic-liquid-crystal-aerogel system, *Phys. Rev. E* 55: 504–507.

[36] Mertelj, A., Spindler, L., and Čopič, M. (1997). Dynamic light scattering in polymer-dispersed liquid crystals, *Phys. Rev. E* 56:549–553.

[37] Čopič, M. and Mertelj, A. (1998). Reorientation in random potential: A model for glasslike dynamics in confined liquid crystals, *Phys. Rev. Lett.* 80:1449–1452.

[38] Jazbinšek, M., Drevenšek Olenik, I., Zgonik, M., Fontecchio, A.K., and Crawford, G.P. (2001). Characterization of holographic polymer dispersed liquid crystal transmission gratings, *J. Appl. Phys.* 90:3831–3837.

[39] Jazbinšek, M., Drevenšek Olenik, I., Zgonik, M., Fontecchio, A.K., and Crawford, G.P. (2002). Electro-optical properties of polymer dispersed liquid crystal transmission gratings, *Mol. Cryst. Liq. Cryst.* 375:455–465.

[40] Coles, H.J. and Bancroft, M.S. (1993). Viscosity coefficients and elastic-constants of nematic solutions of a side-chain polymer, *Mol. Cryst. Liq. Cryst.* 237:97–110.

[41] De Sarkar, M., Qi, J., and Crawford, G.P. (2002). Influence of partial matrix fluorination on morphology and performance of HPDLC transmission gratings, *Polymer* 43:7335–7344.

[42] De Sarkar, M., Gill, N.L., Whitehead, J.B., and Crawford, G.P. (2003). Effect of monomer functionality on the morphology and performance of the holographic transmission gratings recorded on polymer dispersed liquid crystals, *Macromolecules* 36:630–638.

[43] Vilfan, M., Zalar, B., Fontecchio, A.K., Vilfan, M., Escuti, M.J., Crawford, G.P., and Žumer, S. (2002). Deuteron NMR study of molecular ordering in a holographic-polymer-dispersed liquid crystal, *Phys. Rev. E* 66:021710(1–9).

[44] Amundson, K. (1996). Electro-optic properties of a polymer-dispersed liquid-crystal film: Temperature dependence and phase behavior, *Phys. Rev. E* 53:2412–2422.

[45] Yokoyama, H., Kobayashi, S., and Kamei, H. (1984). Deformations of a planar nematic-isotropic interface in uniform and nonuniform electric-fields, *Mol. Cryst. Liq. Cryst.* 129:109–126.

[46] Bharadwaj, R.K., Bunning, T.J., and Farmer, B.L. (2000). A mesoscale modelling study of nematic liquid crystals confined to ellipsoidal domains, *Liq. Cryst.* 27: 591–603.

Chapter 5

TIME-RESOLVED FLUORESCENCE AND DICHROISM IN ABSORBING LIQUIDS

Intermolecular Guest–Host Interactions in Absorbing Liquids

D. Paparo, C. Manzo, and L. Marrucci

Abstract   A detailed understanding of guest–host molecular interactions is required in many topical research fields ranging from biology to the physics of photosensitive materials. Recently, we proved that time-resolved and polarization-sensitive fluorescence and dichroism are effective experimental tools for elucidating the influence of specific intermolecular interactions on the nonlinear optics of anthraquinone-doped liquids and liquid crystals. We review these results in this chapter. In particular, we report on the application of fluorescence and dichroism to the study of the influence of isotope deuterium substitution and photoinduced electronic excitation on the rotational dynamics of anthraquinone molecules in liquids. Our results point to photosensitive hydrogen bonding as the main microscopic mechanism for the large enhancement of optical nonlinearity observed in these materials. Moreover they prove the feasibility of a 'fluctuating-friction molecular motor', a concept which could have wide-ranging implications for controlling the state of matter at the molecular scale. This proof of concept relies on the accurate characterization of the state-dependent rotational dynamics of anthraquinone molecules. To this aim we developed an experimental methodology that combines measurements of transient bleaching and of time-resolved fluorescence and spectroscopy. This approach appears to be the only viable one whenever light absorption of the dye excited states cannot be made negligible.

5.1   Introduction

Time-resolved (TR) and polarization-sensitive (PS) optical spectroscopies based on fluorescence and dichroism have proved to be very effective for studying the rotational dynamics of anisotropic molecules in a wide variety of solvents. Several experimental schemes have been proposed so far in the literature to investigate different features of the molecular dynamics [1, 2].

In this chapter, we do not describe the development of a novel experimental scheme for TRPS fluorescence and dichroism measurements. Indeed, the experiments we review have been performed by means of standard setups as described, for example, in [1, 2]. Instead the novelty of our approach is in the specific way of combining different information from TRPS fluorescence and dichroism to achieve a complete characterization of guest–host photosensitive intermolecular interactions in anthraquinone-doped liquids and liquid crystals. Our interest in these materials was motivated by the search for an explanation of the anomalous nonlinear optical effects observed in dye-doped liquids [3, 4] and liquid crystals [5]. In the first section of this chapter, we briefly introduce the photoinduced molecular phenomena believed to lead to the anomalous nonlinear optical response of anthraquinone-doped liquids and liquid crystals. In particular, we discuss some experimental results that highlight the role of specific intermolecular interactions in these photoinduced effects. These experimental findings demonstrate a strong sensitivity of the anomalous optical nonlinearity to chemical and deuterium-isotope substitution. Next, we show that the anomalies observed in the nonlinear optical response of anthraquinone-doped liquids and liquid crystals have a correspondence in the anomalous rotational dynamics of these dye molecules in liquid solutions. In particular, we focus our attention on two experiments demonstrating, respectively, the influence on this rotational dynamics of deuterium-isotope substitution and photoinduced electronic excitation. In the first, we measure a reduced rotational mobility of the deuterated species with respect to the protonated ones by up to 43% (at 311 K). To our knowledge, this is the largest isotope effect on rotational mobility ever reported in literature. In the second experiment, we find a substantial decrease of the rotational mobility by 30–50% when the dye molecule is promoted from the ground state to its first excited singlet state. We show that both these results quantitatively explain the correspondent observed anomalies in the optical nonlinearity of these systems.

These experimental findings on the rotational dynamics of anthraquinone dyes in liquid hosts are also interesting as they show deviations from hydrodynamic theories usually used for describing the rotational diffusion of dye molecules in liquid solvents [6–12]. These theories are commonly gathered in the literature under the name of Stokes–Einstein–Debye theory (SED). SED models treat the solvent as a macroscopic continuum, in which the diffusional rotation of the solute is only affected by the viscosity and temperature of the hosting solvent. However, as expected, the validity of this continuum description breaks down when the size of the solute molecules approaches that of the solvent molecules or becomes smaller. In this regime that includes the important case of pure materials, the specific intermolecular interactions between the solute and the solvent molecules start to play a fundamental role in their rotational dynamics [9, 11–16].

Despite the failure of SED models, anomalous behaviors in the rotational diffusion of solute molecules provide us with the interesting opportunity of using TRPS fluorescence and dichroism for obtaining microscopic information on the specific intermolecular interactions in complex liquids as we show in the following.

This chapter has the following structure. In Sect. 5.2, we briefly summarize the nonlinear optics of dye-doped liquids and liquid crystals. In particular we discuss some experimental results that highlight the role of specific intermolecular interactions in the nonlinear optical properties of these systems. Section 5.3 is devoted to some theoretical details of TRPS fluorescence and dichroism. Sections 5.4 and 5.5 review our recent experimental results on fluorescence and dichroism in absorbing liquids. According to the focus of this book, we describe those experimental details necessary in achieving accurate quantitative results. These results point to hydrogen bonding as the microscopic mechanism acting in the nonlinear optics of anthraquinone-doped liquids and liquid crystals. Therefore in the conclusive section (Sect. 5.6) we present an extensive discussion of our results within the framework of hydrogen-bonding interactions.

5.2 Photoinduced Molecular Reorientation of Absorbing Liquids and Liquid Crystals

The electric field of a light beam exerts a torque on anisotropic molecules by coupling to the oscillating dipole induced in the molecule by the field itself. The resulting light-induced molecule reorientation is the main mechanism for optical nonlinearity in transparent liquids, the so-called optical Kerr effect [17].

In light-absorbing liquids, other phenomena, such as thermal lensing and incoherent electronic excitation, may contribute to the material nonlinearity. However, the phenomenon of light-induced molecular orientation itself can be strongly modified when the liquid is doped with small quantities of certain dyes. In particular, their nonlinear birefringence resulting from light-induced molecular orientation may be enhanced by orders of magnitude [3, 4, 18]. Although still related to molecular reorientation, this effect cannot be explained by a simple enhancement of the optical dipole torque. Its explanation instead must consider the transformations occurring in dye molecules as a result of photoinduced electronic excitation and the consequently altered equilibrium of molecular orientational interactions.

In the last decade detailed theoretical works have provided us with a satisfactory explanation of this effect (see [19] and references therein). So far two possible mechanisms for the photoinduced enhancement of optical Kerr effect have been considered. They are shown pictorially in Fig. 5.1, where for simplicity we consider dichroic anisotropic molecules with a dipole moment

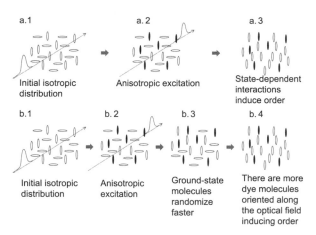

Fig. 5.1 Molecular mechanisms of photoinduced reorientation (see text for explanation). For simplicity we have drawn only two orientations. Note that we have indicated with the same symbol both host and ground-state dye molecules (*open ellipsoids*), while distinguishing photoexcited dye molecules (*filled ellipsoids*)

along the major molecular axis. The first mechanism relies on a photoinduced change of the equilibrium interactions between the solvent and solute molecules (panel a of Fig. 5.1). Initially the orientational distribution for both dye and host molecules is isotropic (for simplicity only two orientations have been indicated in the figure). Then, a linearly polarized optical pulse excites some of those molecules aligned preferentially along the optical field. If dye molecules interact differently with the host molecules depending on their electronic state, this difference induces an effective molecular mean field that breaks the initial isotropic symmetry and forces the reorientation of the host molecules (we are neglecting here the direct contribution of the optical field to the molecular reorientation). In the example shown in figure, the excited dye molecules interact stronger with the host molecules than those in the ground state. This forces a reorientation of the host molecules toward the optical field. On the other hand, if the ground-state molecules interact stronger, the host molecules are reoriented toward a direction perpendicular to the optical field. The latter behavior is absolutely anomalous for the standard optical Kerr effect acting in transparent liquids and is conventionally indicated as "negative" effect. It has been recently demonstrated in azo-doped glycerin [18].

Alternatively, in the absence of a photoinduced change of the equilibrium interaction potential, the same enhancement may be caused by a photoinduced variation of the dye rotational mobility. This mechanism is described in panel b of Fig. 5.1. The example in figure assumes that, after photoexcitation, the orientation of dye molecules in the ground state randomizes much faster than in the excited state. As a consequence, there is an accumulation of dye molecules oriented along the optical field that in turn may force the host molecules to reorient toward this direction. Also this second model can predict a "negative" photoinduced effect that occurs when the rotational diffusion of the excited-state molecules is faster than that of the ground-state molecules.

Actually, the idea of searching for a dye-induced enhancement of the optical Kerr nonlinearity in absorbing liquids was triggered by the discovery of an analogous phenomenon in the nematic phase of dye-doped liquid crystals [5]. Apart from the complications associated with the collective response of the nematic phase, the possible driving mechanisms for this phenomenon are analogous to those described in Fig. 5.1 (see [20]). The efficiency of a particular dye in reorienting the liquid or liquid-crystal host is measured by a merit figure, μ, first introduced in [21], which gives an estimate of the average orientational action that each dye molecule exerts on the host if it absorbs a single photon. We refer the reader to this paper and references therein for a detailed discussion on it. Nevertheless it is useful to write here its analytical expression in the limit of low host anisotropy to gain an insight into the role played by the microscopic parameters in the effect. It can be demonstrated that the following relation holds [20]

$$\mu - \frac{2}{15h}\tau_d u_g \left(\frac{u_e}{u_g} - \frac{D_e}{D_g}\right), \qquad (5.1)$$

where $\tau_d^{-1} = \tau_c^{-1} + 6D_e$, τ_e is the dye excited-state lifetime, D_e (D_g) the rotational diffusion constant for the excited (ground) state, h the Planck's constant, and u_e (u_g) the mean-field intermolecular interaction potential for the excited (ground) state. A merit figure $\mu \approx 3,500$ corresponds to an enhancement of the optical nonlinearity of two orders of magnitude, while a negative value of μ corresponds to a "negative" photoinduced effect. Apart from the other microscopic constants, (5.1) clearly shows that the sign and strength of μ depend on the relative weight of the ratios D_g/D_e and u_g/u_e. The two mechanisms described in Fig. 5.1a and b correspond to $u_e/u_g > 1$ and $D_g/D_e > 1$, respectively, and (5.1) allows for their simultaneous action. The nonlinearity itself, therefore, cannot distinguish between these two contributions. The only possibility of discriminating between them is to measure directly the ratios u_e/u_g and D_g/D_e. As we see in Sect. 5.5, this can be accomplished by combining TRPS fluorescence and dichroism spectroscopies.

Among all the experimental investigations performed in anthraquinone-doped nematic liquid crystals, two results are worth being noted here as they highlight some important features of the microscopic intermolecular interactions involved in this phenomenon. In the first experiment, we have found that slight changes in the molecular structure of relatively simple anthraquinone dyes may lead to dramatic modifications of the macroscopic nonlinear optical properties of the liquid-crystalline host [22]. This is shown in Fig. 5.2

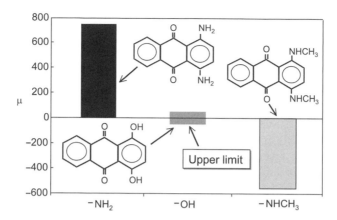

Fig. 5.2 Merit figure μ of three guest–host systems, each composed of the same nematic mixture of biphenyls (E7) and one of the three dyes shown in figure (dye concentration is about 10^{-3} in weight)

where the merit figure μ is reported as a function of the dye substituents. This figure clearly shows that an anthraquinone dye substituted with NH_2, OH, and $NHCH_3$ exhibits a strong positive, vanishing, and strong negative enhancement of the optical nonlinearity, respectively. This dependence on the dye functional groups is also affected by the specific liquid-crystalline host. In particular, it is strong in those liquid crystals mainly formed by molecules that have proton acceptor groups, such as the –CN cyano group, while it is negligibly small in liquid crystals where such groups are absent.

In the second experiment we modified the molecular structure by isotopically substituting hydrogens with deuterium atoms in the functional groups of certain dyes [23]. Unexpectedly, these extremely tiny changes led to an approximate doubling of the magnitude of the already strong optical nonlinearity of the dye-doped nematic liquid crystals. The details of the materials used in this experiment will be given in Sects. 4 and 5.5 where we show that this enhancement is well correlated with a corresponding variation of the fluorescence lifetime and rotational mobility of the dye. It is enough to point out here that the main molecular characteristic of these dyes is the presence of amino ($-NH_2$) radicals in their structure. In particular, this is the only substituent group present in one of the dyes showing the strongest deuterium effect. Also in this experiment, we found that the strongest deuterium effect occurs in liquid crystals with proton-acceptor groups in their molecules. The occurrence of these effects in dye-doped liquid crystals with these molecular characteristics strongly suggests that specific polar interactions, most likely hydrogen bonding, are at the root of the photoinduced enhancement of optical nonlinearity in anthraquinone-doped liquids and liquid crystals. We discuss this point in Sect. 5.6.

5.3 Theory

In this section, we briefly outline some theoretical concepts useful for analyzing the experimental results of fluorescence and dichroism reported in Sects. 4 and 5.5.

We consider molecules that are elongated along one dimension. Experimentally we find that their rotational motion may be fully described by neglecting any rotation around their major axis of symmetry (at least within the timescale of our experiments). Therefore, we assume that our molecule may be schematized as a segment. This hypothesis is known in literature as "rod-like" approximation [24].

Now let us consider the equations that describe the rotational diffusion of rod-like dye molecules in liquid hosts. We restrict our attention to dilute solutions in which dye–dye interactions may be neglected. Moreover, we neglect any photoinduced variation of the orientational distribution of host molecules. This approximation is valid in the limit of low pump intensities. We consider different electronic states of the same dye, which we label as different i-type molecular populations. The instantaneous number of i-type molecules per unit volume and solid angle is given by $f_i(s_i, t)$, where s_i is a unit vector specifying the direction of the molecule long axis. Under the influence of pumping light these functions change in time due to photoinduced excitations and subsequent deexcitations. Besides the photoinduced electronic transitions, the molecules are subject to rotational diffusion (i.e., Brownian motion in the angular space). In the so-called diffusional approximation, corresponding to the assumption that the reorientation of a molecule occurs in small angular steps, all these processes can be combined in the following set of dynamical rate equations written in polar coordinates [24, 25]:

$$\frac{\partial f_i(\theta, t)}{\partial t} - \frac{D_i}{\sin\theta} \frac{\partial}{\partial\theta} \left[\sin\theta \frac{\partial f_i(\theta, t)}{\partial\theta} \right] \sum_{j\neq i} [p_{ji}(\theta, t)f_j - p_{ij}(\theta, t)f_i], \quad (5.2)$$

where $p_{ij}(\theta, t)$ is the probability per unit time of having a transition $i \rightarrow j$ in a dye molecule and D_i is the diffusion constant for the dye in the electronic state i. In the next sections, we specify the p_{ij} functions for each experimental geometry. In (5.2), we have assumed that f_i is independent of the azimuthal angle (this statement will be justified in the following).

It is convenient to expand the distribution functions in a series of Legendre polynomials P_l of $\cos\theta$ to give

$$f_i(\theta, t) = \frac{1}{4\pi} \sum_{l=0,2,4,\ldots} (2l+1)Q_i^{(l)}(t)P_l(\cos\theta), \quad (5.3)$$

where

$$Q_i^{(l)}(t) = \int d\Omega f_i(\theta, t) P_l(\cos\theta) \quad (5.4)$$

are the Legendre moments of the distribution. In particular, the zeroth-order moment $Q_i^{(0)} = N_i$ and the ratio $Q_i^{(2)}/N_i$ represent the total number and the orientational order parameter of each population, respectively. By projecting (5.2) on the Legendre basis we obtain the following set of equations for the moments

$$\dot{Q}_i^{(k)}(t) + k(k+1)D_i Q_i^{(k)}(t) = R_i^{(k)}(t), \tag{5.5}$$

where

$$R_i^{(k)} = \frac{1}{4\pi} \sum_{j \neq i} \sum_l (2l+1) \left[Q_j^{(l)} \int d\Omega \, p_{ji} P_l P_k - Q_i^{(l)} \int d\Omega \, p_{ij} P_l P_k \right]. \tag{5.6}$$

5.3.1   TRPS fluorescence

Let us first introduce some general definitions that will be useful in the following. We indicate with $\boldsymbol{\mu}_i$ the transition dipole moment relative to the dye in the electronic state i. β_i is the angle between $\boldsymbol{s}_i$ and $\boldsymbol{\mu}_i$, assumed to be constant in time. Thus, we neglect all those molecular internal motions that may modify it. We introduce also the constants $A_i = (3\cos^2\beta_i - 1)/2$. Only linearly polarized pump pulses are considered. We assume that their polarization is parallel to the $\hat{z}$-axis of the reference system. A unit vector $\boldsymbol{u}$ indicates the analysis polarization for both the fluorescent light and the dichroism probe beam. Concerning the spatial intensity profile, the optical pulses used in our experiments may be generally approximated with gaussian beams. However the transverse profile of their intensity is not made explicit in our calculations. Therefore our results are understood as averaged over the beam transverse section. In fluorescence experiments, by using low pumping intensities and by suitably tuning the wavelength, it is always possible to restrict the attention to a given radiative transition from the first excited singlet state (S_1) to the singlet ground state (S_0). This means that we neglect any two-photon process. This allows considering only one i-type population of excited dye molecules labeled with the index $i =$ e.

If $\boldsymbol{\mu}_e \parallel \boldsymbol{s}_e$, it is possible to neglect any rotation around $\boldsymbol{s}_e$ that, in this case, cannot be detected by TRPS fluorescence experiments. The rotation around $\boldsymbol{s}_e$ may be approximately neglected also for dipole moments $\boldsymbol{\mu}_e$ forming with $\boldsymbol{s}_e$ an angle different from zero as long as the rotational dynamics around the major axis is much faster than the temporal resolution of the detection system. Under this hypothesis it may be assumed that the detected fluorescence light is always the result of a temporal average with respect to the rotational motion around $\boldsymbol{s}_e$. This is our assumption here and in the following.

The contribution to the detected fluorescent light from a single dipole $\boldsymbol{\mu}_e$ is given by

$$I(\boldsymbol{\mu}_e) = K(\boldsymbol{u} \cdot \boldsymbol{\mu}_e)^2, \tag{5.7}$$

where K is a constant of proportionality independent of $\boldsymbol{\mu}_e$, whereas the contribution of each molecule averaged around its axis $\boldsymbol{s}_e$ is given by

$$\bar{I}(\boldsymbol{s}_e) = K \left[\frac{1}{3}(1 - A_e) + A_e(\boldsymbol{u} \cdot \boldsymbol{s}_e)^2 \right]. \tag{5.8}$$

Hence, for the fluorescence light detected from all the dye molecules, we are left with the following expression

$$I(t, \boldsymbol{u}) = \int d\Omega \, \bar{I}(\boldsymbol{s}_e) f_e(\boldsymbol{s}_e, t), \tag{5.9}$$

where the integration is over all the solid angle spanned by $\boldsymbol{s}_e$.

By expanding $\bar{I}(\boldsymbol{s}_e)$ in the Legendre basis and by using (5.3), we obtain

$$I(t, \gamma) = K \left[\frac{N_e(t)}{3} + A_e \left(\cos^2 \gamma - \frac{1}{3} \right) Q_e^{(2)}(t) \right], \tag{5.10}$$

where γ is the angle that $\boldsymbol{u}$ forms with the $\hat{z}$-axis. At this point, (5.5) must be solved for $N_e(t)$ and $Q_e^{(2)}(t)$. We divide the solution in two temporal intervals: during and after pump excitation.

During the first interval a pumping light pulse of peak intensity I_p and frequency ν_p induces a probability p_{ge} of promoting the dye from the ground to the excited state given by

$$p_{ge}(\theta) = \frac{\sigma_g(\nu_p)I_p}{h\nu_p} \left[\frac{1}{3}(1 - A_g) + A_g \cos^2 \theta \right], \tag{5.11}$$

where σ_g is the ground-state cross section for light absorption.

We assume that the duration of the optical pulse is so short that diffusion and spontaneous decay of the dye molecule may be neglected during its passage. With these assumptions and after straightforward calculations, it is found that

$$N_e(0^+) = \frac{\sigma_g I_p}{3h\nu} Q_g^{(0)}(0) = \frac{\sigma_g I_p}{h\nu} \frac{N_d}{3} \tag{5.12}$$

$$Q_e^{(2)}(0^+) = \frac{\sigma_g I_p}{h\nu} \frac{2A_g}{15} Q_g^{(0)}(0) = \frac{\sigma_g I_p}{h\nu} \frac{2A_g N_d}{15}, \tag{5.13}$$

where N_d is the total number of dye and 0^+ denotes a time immediately following the pulse passage.

After excitation, molecules diffuse in the absence of the pump field. All the p_{ei} are independent of θ. Therefore $R_e^{(k)} = -Q_e^{(k)}/\tau_e$, where $\tau_e = 1/\sum_{i \neq e} p_{ei}$

is the fluorescence lifetime. The solution of (5.5) with this expression for $R_e^{(k)}$ leads to the following equation for the fluorescence intensity

$$
\begin{aligned}
I(t, \gamma) &= K N_e(0^+) \left[\frac{1}{3} e^{-t/\tau_e} + A_e \frac{Q_e^{(2)}(0^+)}{N_e(0^+)} \left(\cos^2 \gamma - \frac{1}{3} \right) e^{-t/\tau_r} \right] \\
&= K \frac{N_d}{9} \frac{\sigma_g(\nu_p) I_p}{h \nu_p} e^{-t/\tau_e} \left[1 + r_0 \left(3 \cos^2 \gamma - 1 \right) e^{-t/\tau_d} \right], \quad (5.14)
\end{aligned}
$$

where we set $\tau_r = 1/6 D_e$ and $r_0 = (2/5) A_e A_g$. r_0 is independent of dye concentration and pump intensity and gauges the degree of initial excited-dye orientational anisotropy immediately after the pump pulse.

It is interesting to note that there exists an angle $\gamma = \gamma_m \approx 54.7°$, known in literature as "magic angle" [1], for which the fluorescence intensity decay is single exponential with a characteristic time given by the lifetime τ_e. We also note that $I(\gamma_m) = [I(\gamma = 0°) + 2I(\gamma = 90°)]/3 = (I_\parallel + 2I_\perp)/3$. On the other hand, the combination $I_\parallel - I_\perp$ is also a single exponential decay, with characteristic time τ_d.

Expression (5.13) would give the measured signal in case of an infinitely fast response of the detection system. In practice the measured signal $S(t, \gamma)$ is the result of the convolution between the fluorescence intensity $I(t, \gamma)$ and the response function $G(t, \gamma)$ of the detection system, i.e.,

$$
S(t, \gamma) = \int dt' G(t - t', \gamma) I(t', \gamma). \quad (5.15)
$$

5.3.2   TRPS Pump-Probe Dichroism

In contrast to fluorescence, which probes molecules in the excited state only, TRPS dichroism may provide us with information about the dye molecular dynamics in the ground singlet state S_0. In the pump-probe scheme, a linearly polarized pumping optical pulse induces an initial anisotropy in the ground-state population labeled with the suffix $i = g$. Then the decay of this anisotropy is followed by measuring at variable temporal delays the absorption of a probe beam of frequency ν_{pr}.

Dichroism may be described by using the same formalism developed for fluorescence. However, in fluorescence experiments, it is usually possible to restrict the attention to only one electronic state. In general, this is not true for dichroism since the probe transmitted intensity may vary also for the absorption from the excited states that have been previously populated by the pump. Besides the fluorescent state S_1, these excited states usually include also a long-lived triplet state. Therefore, here, we slightly generalize the model used in Sect. 3.3.1 for taking into account more dye electronic states. In particular, we introduce the intersystem crossing rate p_{eT} that describes nonradiative coupling between the S_1 singlet state and the lowest T_1 triplet state (labeled with the index T) and the cross section $\sigma_e(\nu_{pr})$ and $\sigma_T(\nu_{pr})$ for (probe) light absorption from the excited S_1 and T_1 states, respectively.

In our model we assume that vibrational transitions, and deexcitations from higher lying singlet and triplet excited states are so fast that they may be neglected within the temporal scale of our experiment. Moreover the assumption of a large Stokes shift allows neglecting stimulated emission. We also assume that two-photon fluorescence is negligible, as well as two-photon, excited-state, and triplet-state absorption for the pump beam. Finally, we assume that all electronic transitions occur with no change of molecular orientation. We caution that some dyes may possess a more complex photophysics, but it turns out that these assumptions are adequate to a description of the experiments presented in Sect. 5.5. As we did for fluorescence, we assume also that pump-pulse energies are small. Under all these hypothesis, by specializing $R_i^{(k)}$ of (5.5) for each electronic level, it is straightforward to demonstrate that the rotational diffusion of the dye populations is described by the following set of equations

$$\dot{N}_g = p_{eg}N_e + p_{Tg}N_T - \frac{\sigma_g(\nu_p)I_p}{3h\nu_p}\left[N_g + 2A_gQ_g^{(2)}\right], \qquad (5.16)$$

$$\dot{N}_g = -p_{eg}N_e - p_{eT}N_T + \frac{\sigma_g(\nu_p)I_p}{3h\nu_p}\left[N_g + 2A_gQ_g^{(2)}\right], \qquad (5.17)$$

$$N_T = N_d - N_e - N_g, \qquad (5.18)$$

$$\dot{Q}_g^{(2)} = -6D_gQ_g^{(2)} + p_{eg}Q_e^{(2)} + p_{Tg}Q_T^{(2)} \qquad (5.19)$$
$$- \frac{\sigma_g(\nu_p)I_p}{3h\nu_p}\left[\frac{2}{5}A_gN_g + \left(1 + \frac{4}{7}A_g\right)Q_g^{(2)}\right],$$

$$\dot{Q}_e^{(2)} = -6D_eQ_e^{(2)} - (p_{eg} + p_{eT})Q_e^{(2)} \qquad (5.20)$$
$$+ \frac{\sigma_g(\nu_p)I_p}{3h\nu_p}\left[\frac{2}{5}A_gN_g + \left(1 + \frac{4}{7}A_g\right)Q_g^{(2)}\right],$$

$$\dot{Q}_T^{(2)} = -6D_TQ_T^{(2)} + p_{eT}Q_e^{(2)} - p_{Te}Q_T^{(2)}, \qquad (5.21)$$

where we have neglected all the moments having $l > 2$, which do not contribute significantly for low pump intensities. Note that optical transitions between singlet and triplet states are forbidden and that $1/\tau_e = p_{eg} + p_{eT}$. For describing the time-dependent transmission of the probe beam we must calculate the absorption coefficients for the different electronic levels. The overall probe absorption coefficient is given by

$$\alpha(t, \gamma) = \sum_{i=g,e,T} \frac{\sigma_i(\nu_{pr})}{3}\left[N_i(t) + A_i(3\cos^2\gamma - 1)Q_i^{(2)}(t)\right]. \qquad (5.22)$$

The dichroism is usually defined as $\Delta\alpha(t) = \alpha(t, 0°) - \alpha(t, 90°)$. The absorption coefficient given in (5.22) is related to the probe transmittance that is the quantity usually measured in dichroism experiments by the following equation

$$T(t, \gamma) = \exp\left[-\int_0^L \alpha(t, \gamma, l')dl' \right],   \qquad (5.23)$$

where the integration in the exponent is performed along the sample thickness. Note that the dependence of α on the sample thickness is due to the intensity depletion of the pump traveling in the sample. In (5.23), it is also assumed that the probe polarization direction, given by the angle γ, is constant within the sample.

5.4   Fluorescence Experiments: Observing the Molecule Rotation in its Excited State

In this section, we describe two of our recent experiments where steady-state and TRPS fluorescence measurements have revealed many interesting features of the microscopic intermolecular interactions acting in the photoinduced effects described in Sect. 5.2.

For the steady-state fluorescence measurements we used a Varian Cary Eclipse fluorimeter. In the TRPS fluorescence experiments, fluorescent light was induced by pulses generated by a laser system composed of a frequency-tripled Nd:Yag laser pumping an optical parametric generator. This laser system allows generation of pulses having a duration of about 20 ps and with a tunable wavelength within the range 0.4–2 μm. The transient fluorescence signal $S(t, \gamma)$ with $\gamma = 0°, 54.7°$, and $90°$ was detected by means of a fast photodiode (about 100 ps rise time) and electronics with 2 GHz analog bandwidth. In Fig. 5.3, we show typical examples of the detected signals together with the corresponding fitting curves. The nonexponential behavior seen in both data and fit curves is entirely due to the instrumental response function G. Fits are obtained by using a FFT-based algorithm to convolve the predicted signal with the experimentally determined G. It is worth noting how many details of the signal are well reproduced by the fitting curve over three decades. These accurate fits are achieved only after taking care of several experimental issues. In particular, an accurate measurement of the setup response function G is essential. The latter can be measured by collecting the light scattered from an opaque plate placed at the sample position. Due to equipment limitations, the response function is usually measured with a detection occurring at the same wavelength as that used for exciting the fluorescence in the sample. However, because of the Stokes shift, fluorescent light is emitted in an interval of wavelengths significantly

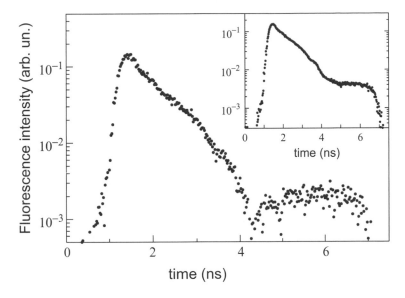

Fig. 5.3 Examples of fluorescence decays in protonated HK271-5CB at a temperature of 42.5°C. In the main figure the temporal decay of the combination $S_{\parallel} - S_{\perp}$ is shown together with the corresponding fitting curve (*gray line*). This signal is actually single exponential with decay time τ_d, but it is convolved with the response function. *Inset*: temporal decay of the combination $S_{\parallel} + 2S_{\perp}$ with relative fitting curve. This behavior is single exponential with decay time τ_e

different from that of the excitation light. On the other hand, the sensitivity and the responsiveness of any detection system is wavelength dependent. As a consequence, the directly measured response function may be in this approach significantly different from the one actually entering the detected fluorescent signal. This difference is a source of systematic error, which should be avoided for improving the fit.

To this purpose, we measured the response function at different wavelengths of the scattered light by exploiting the large tunability of our laser system. Then, by optically filtering the fluorescence signal, we selected a wavelength range within the fluorescence spectrum where the response function was approximately constant. The response function G used in the fitting procedure was then obtained by averaging over all the slightly different response functions measured in this selected wavelength range.

We omit other experimental details that may be found in [16]. According to the theory of Sect. 5.3.1, a best fit of the signal combinations $S_{\parallel} + 2S_{\perp}$ and $S_{\parallel} - S_{\perp}$ provides us with a direct measurement of τ_e and τ_d, and hence of the rotational time τ_r, as well as of the initial anisotropy r_0 (that enters the preexponential factor).

5.4.1   Deuterium Effect on Molecular Rotation and its Relationship with Optical Nonlinearity

In this experiment, we studied the rotational diffusion of two species of amino-substituted anthraquinone dyes dissolved in the isotropic liquid phase of 4'-n-pentyl-4-cyanobiphenyl (5CB) and in a liquid mixture of alkanes (paraffin). The dyes were the 1,8-dihydroxy 4,5-diamino 2,7-diisopentyl anthraquinone (HK271) and the 1-amino anthraquinone (1AAQ). The molecular structures of HK271, 1AAQ, and 5CB are shown in Fig. 5.4. Although 5CB is a liquid crystalline material that has a nematic phase between 297 and 308 K, our study was limited to the temperature range 311–368 K, in which 5CB is in its ordinary isotropic liquid phase. Apart from the motivations outlined in Sect. 5.2, the reason we chose these materials is that they are good candidates for studying possible deviations from simple SED behavior, in connection with the presence of specific interactions. Indeed, the molecule of HK271 is only a factor 1.6 larger in weight than the molecule of 5CB. The molecule of 1AAQ is 10% smaller than that of 5CB. Moreover, as already anticipated in Sect. 5.2, the amino groups of these dyes can hydrogen bond to the cyano group of 5CB (see Fig. 5.4). Since hydrogen bonds are sensitive to isotopic substitution, a significant deuterium effect in the rotational mobility is conceivable [16]. As a reference solvent which cannot hydrogen bond to the dyes, we used a mixture of alkanes $C_n H_{2n+2}$ with $n = 20$–30 (paraffin with a melting point in the range 325–327 K). We prepared also deuterated forms of dyes HK271 and 1AAQ by isotopically substituting with deuterium the hydrogens of amino and hydroxy groups. Other details on the materials as well as on the deuteration procedure may be found in [16].

The main results of this experiment are shown in Figs. 5.5 and 5.6, where the measured values of τ_e and τ_r are reported as a function of temperature for protonated and deuterated solutions HK271–5CB and HK271–paraffin, respectively. The results for 1AAQ–5CB are qualitatively the same of HK271–5CB, and for them we refer the readers to paper [16]. The main figures report, in semiloga-

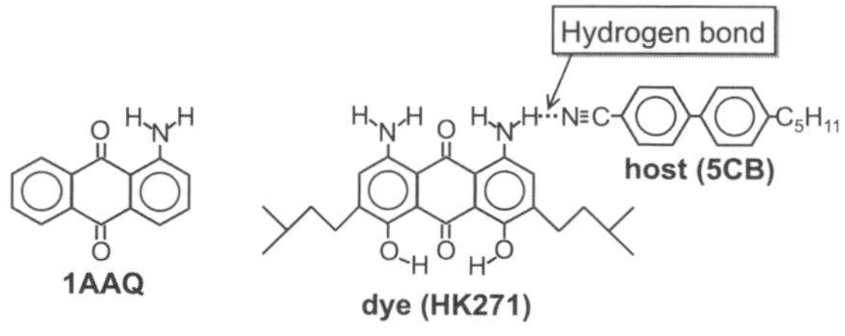

Fig. 5.4 Molecular structures of HK271, 1AAQ, and 5CB. Note the possible intermolecular hydrogen-bond between the dye amino group and the host cyano group (for brevity only the HK271–5CB hydrogen bond is shown)

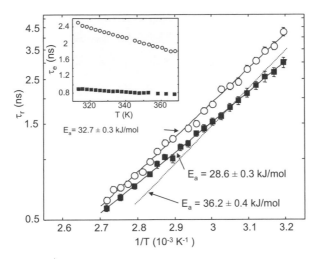

Fig. 5.5 Semilog plot of τ_r vs. $1/T$ for protonated (*filled squares*) and deuterated (*open circles*) HK271–5CB mixture. *Solid lines* are Arrhenius fits. E_a is the resulting activation energy. The *dotted line* is the prediction of SED model for "stick" boundary conditions. *Inset*: fluorescence lifetime τ_e vs. T for the same samples (same meaning of symbols). Reprinted with permission from [16]. Copyright 2002, American Institute of Physics

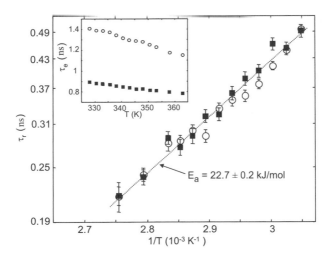

Fig. 5.6 Semilog plot of τ_r vs. $1/T$ for protonated (*filled squares*) and deuterated (*open circles*) HK271–paraffin mixture. The *dotted line* is the prediction of SED model with "slip" boundary conditions. The Arrhenius activation energies of both data sets are consistent with that of SED, given in figure. *Inset*: corresponding fluorescence lifetimes. Reprinted with permission from [16]. Copyright 2002, American Institute of Physics

rithmic scale, the rotational times vs. inverse absolute temperature (error bars are for a confidence level of 99%, not including the statistical uncertainty on the response function). The solid lines are best fits to data based on the phenomenological Arrhenius law $\tau_r = \tau_0 \exp(E_a/kT)$ that describes activated processes, where k is the Boltzmann constant, E_a is the activation energy (best fit values are given in the figures), and τ_0 is a constant. The fluorescence lifetime τ_e of the corresponding samples are plotted in the figure insets.

A very large isotopic effect of both lifetime τ_e and rotational time τ_r in HK271–5CB and of τ_e only in the paraffin solution is evident in our data. The isotopic effect of τ_r is our main result and it is discussed in the following. First we briefly comment on the isotope effect on lifetime τ_e. Similar effects have been already reported for related dyes (see references in [16]). Probably the nonradiative decay channels of state S_1 associated with nuclear vibrations are suppressed by the deuterium substitution. Intermolecular interactions and possibly hydrogen bonding must also play an important role in this phenomenon, as we found that the deuterium effect on τ_e of HK271 is much larger in 5CB than in paraffin. Solvent-dependent nonradiative decay attributed to intermolecular hydrogen bonding has been already reported for anthraquinone dyes (see references in [16]).

Let us turn now to the isotopic effect of the rotational time τ_r (Fig. 5.5). The increase of τ_r induced by deuteration is of about 40% at the lowest investigated temperatures (313 K). To our knowledge, this is the largest isotopic effect of diffusional rotational dynamics ever reported. The isotopic effect on τ_r vanishes completely (within the uncertainties) when HK271 is dissolved in liquid paraffin (Fig. 5.6). This shows unambiguously that specific interactions, most likely hydrogen bonding, between HK271 and 5CB are the main cause of the effect. The comparison with the SED model strongly supports this view. The curves shown in the figures as dotted lines have been calculated on the basis of SED theory and of molecular parameters as calculated in [16]. The comparison clearly shows that SED description breaks down in 5CB solvent due to the strong contribution of the specific intermolecular interactions, while it works perfectly in the case of paraffin (Fig. 5.6). In the latter case SED theory applies with the "slip" boundary conditions, meaning that the solvent exerts no tangential stress on the molecule. This confirms the absence of strong intermolecular interactions for our dye in solvent not capable of hydrogen bonding.

In HK271–5CB the SED predictions shown in Fig. 5.5 (dotted line) are obtained by applying "stick" boundary conditions, corresponding to a zero relative velocity between the spheroid surface and the first fluid layer. We can see that this system is in the so-called super-stick regime, i.e., almost all data lie above the SED line (analogous results are found for the system 1AAQ–5CB). This behavior has no explanation within a pure hydrodynamic model and highlights once more the importance of specific interactions in these systems, as discussed in Sect. 5.6. A detailed discussion on the deviations of our results from the SED

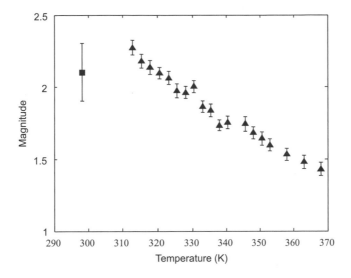

Fig. 5.7 Ratios of the effective lifetime τ_d for deuterated and protonated samples vs. temperature (*triangles*). For comparison, the ratio of the optical nonlinearity magnitudes for the same samples, measured in the nematic phase, is also shown (*square*)

model is outside the scope of this chapter. For further considerations on this point we refer the readers to [16].

From Fig. 5.7, we see that the ratio of the effective lifetime τ_d for deuterated and nondeuterated materials approaches the ratio of the corresponding optical nonlinearity at the lowest temperature. This is in agreement with the predictions of (5.1) if we may also assume that u_g, u_e, and the ratio D_g/D_e are approximately unaffected by the substitution. We verified that u_g is indeed unaffected by measuring the linear dichroism of the dye-liquid crystal solutions and finding it unchanged in the deuterated species (see [21] for the relationship between u_g and dichroism). It is hence reasonable to assume that also u_e is unaffected. Moreover, the ratio D_g/D_e is also not much affected by deuterium substitution as reported in Sect. 5.5. We should also mention that (5.1) is actually exact only in the limit of small liquid crystal order parameter. Nevertheless, a more precise numerical calculation confirms the agreement [23].

5.4.2 Effect of Excitation-Light Wavelength on Dye Excitation and its Relationship with Optical Nonlinearity

In [26], we report a detailed study on TRPS and steady-state fluorescence of HK271 in different liquids as a function of the excitation-light wavelength, λ. These experiments have been motivated by the observed λ-dependence of the merit figure μ in certain dye-doped liquid crystals [27, 28]. In particular, for HK271 in a liquid crystalline mixture of cyanophenyls (E63), μ increases by increasing λ. This dependence is difficult to explain if only the singlet S_1 excited state is involved in the photoinduced enhancement of nonlinear optical properties. It could point to the occurrence in HK271 of different kinds of polymorphism of the dye excited state, such as excitation of long-lived internal vibrations by the photon extra energy, variations in the intramolecular or intermolecular hydrogen bond or other physical interactions (e.g., an excess photon energy may lead to breaking a hydrogen bond), and so on. All these changes, however, should be reflected in a λ-dependence of the microscopic parameters τ_r, τ_e, D_g, u_e, or u_g. The experiments reported in [26] show that this is not the case, as we find no significant variation of τ_r and τ_e by varying λ in the absorption band of the dye. Further considerations lead us to exclude any influence also on the other parameters (see [26] for a detailed discussion on this point).

A second possibility is that the fast vibrational relaxation immediately following the photon absorption, whose magnitude depends on the photon excess energy, could lead to a significant random reorientation of the excited dye molecules (see [28]). The occurrence of this effect should be reflected into a λ-dependence of the initial anisotropy r_0. We have explored also this possibility, finding, however, a substantial invariance with λ of r_0 in HK271–5CB solution. The measurements of the λ-dependence of τ_r, τ_e, D_g, and r_0, may be found in [26]. We are left with a last possible explanation that relies on the hypothesis that a λ-dependent fraction of the excitations do not usefully participate to improve μ. This could be the case, for example, if the dye molecule relaxes nonradiatively back to the ground state immediately after the excitation, or if the excitations are immediately followed by an intersystem crossing to an ineffective triplet state. Whatever its detailed nature, if it occurs, this effect should usually be more likely at higher photon energies (lower λ), where more decay channels are present [29]. This effect should also be reflected into a λ-dependence of the fluorescence quantum yield Φ that monitors the S_1-fluorescent state population induced by absorption. The quantum yield $\Phi(\lambda)$ is obtained by taking the ratio of the number of fluorescence photons as deduced from the excitation steady-state fluorescence spectrum (i.e., fluorescence intensity at a fixed emission wavelength for a varying excitation wavelength) to the number of absorbed photons as measured from the absorption spectrum. Figure 5.8 shows the measured $\Phi(\lambda)$ for HK271 in 5CB, paraffin, and ethanol at room temperature. It is seen that $\Phi(\lambda)$ is significantly nonconstant only in 5CB. A very small increase is perhaps present in ethanol, while $\Phi(\lambda)$ is constant in the nonpolar paraffin. This solvent effect may be related to the presence of host-dependent nonradiative decay channels, in connection with intermolecular hydrogen bonding between HK271 and 5CB.

In Fig. 5.9, the effect of deuteration on the $\Phi(\lambda)$ behavior for HK271 in 5CB and in paraffin is reported. It is seen that deuterated samples exhibit the same

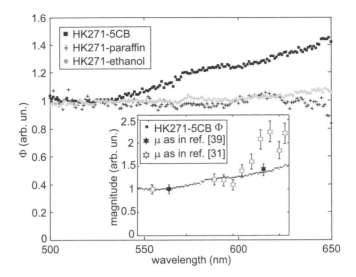

Fig. 5.8 Fluorescence quantum yield Φ vs. excitation wavelength λ for the dye HK271 dissolved in three different hosts, namely 5CB, paraffin, and ethanol. The three Φ curves have been rescaled so as to be superimposed at $\lambda = 532$ nm. *Inset*: $\Phi(\lambda)$ of HK271–5CB compared to $\mu(\lambda)$ of HK271–E63 as measured in two different experiments (the x-axis is the same of the main figure). $\mu(\lambda)$ has been rescaled so as to be superimposed to $\Phi(\lambda)$ at the lowest wavelengths

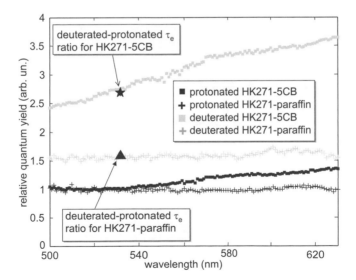

Fig. 5.9 Effect of hydrogen–deuterium substitution on the quantum yield $\Phi(\lambda)$. The quantum yields of the protonated solutions have been rescaled so as to be superimposed at the lowest wavelengths. The data for the deuterated compounds have been kept in scale with the corresponding protonated ones. At $\lambda = 532$ nm, the deuterated-protonated lifetime ratio is also shown for HK271 in 5CB and paraffin

$\Phi(\lambda)$ as protonated ones except for a λ-independent overall multiplicative enhancement. This enhancement is in excellent agreement with the corresponding increase of the excited-state lifetime τ_e measured at $\lambda = 532\,\text{nm}$, as shown in the figure.

In the inset of Fig. 5.8, the correlation between the λ-dependence of Φ for HK271–5CB and the merit figure μ measured in two different experiments is shown. This correlation demonstrates that the anomalous λ-dependence of the optical nonlinearity is due to a decrease of the S_1-state population for increasing excitation photon energy. For further details on this point we refer the reader to [26].

5.5 Combining Fluorescence and Dichroism Experiments: The Influence of Electronic Excitation on Rotational Friction

To assess the contribution of the two mechanisms for optical nonlinearity discussed in Sect. 5.2, we need to measure the ratios D_g/D_e and u_g/u_e, which enter (5.1). To this purpose, we combined the information on the S_1 state obtained by TRPS fluorescence with that of TRPS pump-probe dichroism that mainly probes the dynamics of the S_0 ground state. However, unlike the fluorescence case, the signal obtained by means of the latter technique is often of difficult interpretation. When the dye presents a region of wavelengths where absorption takes place only from the ground state, then probing in this region reveals the relaxation of the ground-state dye population resulting from its own rotational diffusion, as controlled by the constant D_g known from TRPS pump-probe dichroism, and from the relaxation of excited molecules, as determined by the constants τ_e and D_e known from TRPS fluorescence.

In some cases negligible excited-state absorption for some probe wavelengths may be identified (see [30] and references therein). However, in general, one has to face with a complex electronic multilevel system leading to a significant excited-state absorption at all wavelenghts, and a suitable experimental methodology based on the combination of different measurements must be applied [30]. In our experiment, transient absorption was measured using two independent setups. In the first setup, we used a 5-ns pulse at 532 nm to excite the dye and a cw probe at a wavelength adjustable in the range 600–700 nm. We measured the variations of absolute probe transmittance during and after pump passage, varying the probe polarization relative to the pump one, the probe wavelength (to vary the relative degree of ground-state and excited-state absorption), and the pump pulse energy. Examples of our results with this setup are presented in Fig. 5.10. In the second setup, we employed the same laser system described in Sect. 5.4, with 20-ps laser pulses used both for pumping and probing, with an adjustable delay. The dichroism was detected, in this case, in the crossed polarizers geometry, leading to a signal proportional to $\Delta\alpha^2$. An

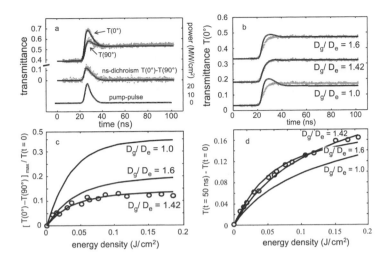

Fig. 5.10 Examples of ns pump-probe absorption spectroscopy for deuterated HK271 in 5CB. They show some indicators of our global best-fit procedure. (**a, b**) Typical probe transmittance signals (*gray noisy line*). The apparently constant isotropic signal reached after the pump passage is actually decaying in a time of several microseconds (data not shown). (**c**) Maximum ns dichroism, normalized to equilibrium transmittance $T(t = 0)$, vs. pump pulse energy density (*circles*). (**d**) Transmittance measured at 50 ns (triplet contribution) vs. pump pulse energy density. Best-fit curves are reported as *black lines* (see text for explanation). Reprinted with permission from [30]. Copyright 2003, American Physical Society

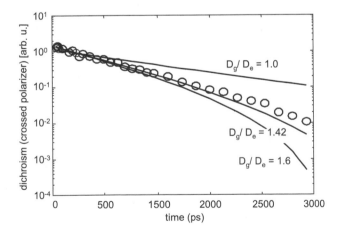

Fig. 5.11 Semilog plot of transient dichroism detected with the pump-probe picosecond set-up vs. time after the pump pulse passage. Note that transmittance was measured in the crossed polarizers geometry

example of this signal is shown in Fig. 5.11. As already stated in Sect. 5.3.2, the main difficulty is to extract information about the ground-state dynamics, while the signal is also determined by the absorption from the S_1 excited state and from the lowest triplet state T_1. To overcome this problem, we exploited the additional information that can be gained by measuring the absorbance $\alpha(t)$ of the probe induced in the sample by the pump light as a function of its intensity. The long-term (time scale of microsecond; see also Fig. 5.10a,b) behavior of $\alpha(t)$ is indeed sensitive to the triplet state only, and its intensity-dependence (Fig. 5.10d) allows for a determination of the triplet cross section $\sigma_T(\nu_{pr})$ and intersystem-crossing rate p_{eT}. The inverse of the T_1-state lifetime gives p_{Tg}. The intensity dependence of the short-term (ns) absorbance signal $T(\gamma = 0°) - T(\gamma = 90°)$ (Fig. 5.10c) is instead mainly sensitive to the S_1-state cross section $\sigma_e(\nu_{pr})$. Finally the fast dichroism data shown in Fig. 5.11 is mainly sensitive to D_g and $\sigma_e(\nu_{pr})$. Therefore, for each material, temperature, and probe wavelength, we performed a global fit on all relevant data (ns-probe absorbance for both polarizations vs. time and pump energy, and ps-dichroism), using the system of equations (5.16)–(5.21) as model. The solution of the latter was obtained numerically. We adjusted the four parameters D_g, p_{eT}, $\sigma_e(\nu_{pr})$, and $\sigma_T(\nu_{pr})$ for obtaining the best fit to all data. The parameters D_e and τ_e were kept fixed to the values obtained from our fluorescence measurements, whereas $\sigma_g(\nu_{pr})$ was determined from the equilibrium absorbance. We verified that an A_e very different from 1 did not allow for good fitting, so we assumed $A_e = A_T = 1$. The best-fit results for D_g were found to be robust and quite insensitive to many details, including the weights assigned to different data sets, and the value given to the constant D_T. Fixing D_T to any value in the range $[D_e, D_g]$ yielded the same results within 1%. Examples of our best-fit curves are shown in Figs. 5.10 and 5.11 obtained for $D_g/D_e = 1.42$. The other two curves are obtained by fixing D_g/D_e, respectively, to 1 and 1.6 and adjusting the other parameters for best fit. The results clearly show the reliability of our estimate of D_g/D_e.

The final results obtained for the ratio D_g/D_e as a function of temperature and material are shown in Fig. 5.12. We see that in all investigated samples the rotational diffusion constant in the excited state D_e is consistently smaller than the ground-state one D_g by 30–50%. This result is roughly independent of temperature, within measurements errors. Deuteration is found to enhance only slightly the mobility ratio. The photoinduced reorientation described in Sect. 5.2 can be explained quantitatively by the model reported in [25] for a ratio $D_g/D_e = 1.4$, that is fully consistent with the result of our direct measurement. This would imply that $u_e/u_g \simeq 1$. However it is also possible that u_e/u_g is somewhat larger than unity and that this additional contribution is compensated by the losses discussed in Sect. 5.4.2, in connection with the λ-dependence of μ. The doubled reorientation efficiency of deuterated HK271 in 5CB and E63 is entirely due to the doubling of the excited-state correlation time τ_d that gauges the duration of the photoexcited dye effect (see also Sect. 5.4 and, for a more detailed discussion on this point, [23]). Therefore, the

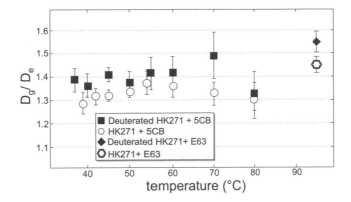

Fig. 5.12 Ground-state to excited-state rotational mobility ratio D_g/D_e for different guest–host systems vs. sample temperature. Error bars are given for a fit confidence level of 99%. Reprinted with permission from [30]. Copyright 2003, American Physical Society

expected ratio D_g/D_e in deuterated HK271 solutions is approximately the same as in nondeuterated ones. Again this is confirmed by the results of our direct measurements. This agreement shows conclusively that the main underlying driving force of photoinduced reorientation occurring in these materials is the light-induced modulation of dye rotational mobility.

5.6 Conclusions: The Role of Hydrogen Bonding

The experiments described in Sects. 5.4 and 5.5 show that it is possible, by combining different spectroscopic techniques namely TRPS fluorescence and dichroism, to gain a good understanding of the microscopic intermolecular interactions acting in certain guest–host systems. All the results strongly support the idea that in the excited state S_1 of anthraquinone dyes with amino substituents, there is a strong reinforcement of the hydrogen bonds between the amino groups and the proton-accepting groups of the solvent (see Fig. 5.4). As shown in Fig. 5.13, this reinforcement is most likely associated with an intramolecular charge-transfer consequent to the dye electronic excitation [31,32].

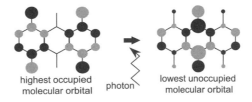

Fig. 5.13 Results of a linear-combination-of-atomic-orbitals calculation of the effect of photoexcitation on π-electron distribution in 1,4,5,8-tetramino anthraquinone. The size of the *black* (*gray*) *circles* gives the positive (negative) wavefunction amplitude for the electrons localized in the p_z-orbitals [31]

Hydrogen-bond strengthening in excited states has been reported before in similar systems [33]. This hypothesis is also supported by the observed dependence of the reorientation effects on the dye substituent groups and host polarity as discussed in Sect. 5.2, and of the excited-state rotational mobility on deuteration as discussed in Sect. 5.4.1. Another confirmation comes from the observed deuterium effect on the characteristic time τ_d that correlates well with the corresponding amplification of the optical nonlinearity in the deuterated samples. In contrast to the singlet excited state S_1, the triplet state T_1 probably does not contribute to the nonlinearity, as it is likely characterized by a weakening of the hydrogen-bonding capability. This picture is also well supported by measurements of acidity of the excited states in amino-substituted anthraquinones reported in [34]. For these dyes, the authors measure acidity values of the T_1 triplet state much closer to the S_0 ground state than to the S_1 excited state. Since acidity is an indirect measure of the dye capability of sharing a proton in a hydrogen bond, these results strongly confirm our experimental findings.

The experimental results of Sect. 5.4.1 disclose also other intriguing aspects of this specific hydrogen bond. Indeed the temperature dependence of τ_r for both dyes in 5CB shows an increase of 4–5 kJmol^{-1} in the activation energy of the deuterated species with respect to the protonated ones. This activation energy that we obtain from our Arrhenius fits is a phenomenological parameter that has no simple microscopic interpretation. Nonetheless, it is interesting to compare our results with the effect of deuteration on the hydrogen-bond equilibrium energy. Typically, this energy change is about 1 kJmol^{-1} or less [35], which is not enough to explain our results. One could ascribe this discrepancy to a cooperative effect of more than one hydrogen bond. However, the dye 1AAQ is unlikely to make more than one intermolecular bond (and in any case no more than two), because one of the two hydrogens of the amino group is probably already involved in an intramolecular hydrogen bond with the carbonyl group of the anthraquinone core. Although HK271 has the possibility of forming more than one intermolecular bonds (but it is unlikely to have more than two), it shows almost the same isotopic change in activation energy. This leads us to associate the observed isotopic effect to the breaking of a single hydrogen bond. Since the isotopic shift of equilibrium energy is not enough to explain our findings, the main contribution to the variation of activation energy must arise from the change of potential barrier for the breaking of the hydrogen bond, perhaps in connection with the diminished quantum delocalization of deuterium [36]. Therefore, the effect of deuteration on the rotational mobility appears to be essentially kinetic in nature, probably associated with an increase of the energy barrier for hydrogen bond breaking upon deuteration. These findings could be relevant for the modeling of a broad class of kinetic phenomena in which hydrogen-bond formation and breaking is a rate-limiting step, including possibly water viscosity [37] and protein folding [38].

Although we have considered only anthraquinone dyes, we point out that our results may be of a more general interest, thus opening several research prospects. First, it is tempting to speculate about the possible role of nonconformational photomobility effect in the photoinduced orientation of azo polymers, currently investigated for their potential applications in optical data storage [39–41], liquid crystals displays [42, 43], and organic optoelectronic devices [44, 45]. Further details on this point may be found in Sect. IV of [30]. From a broader point of view, by proving the molecular reorientation to be caused by a photoinduced modulation of rotational friction, we have also demonstrated the feasibility of a "fluctuating-friction molecular motor" [19, 46]. According to this picture, this microscopic mechanism could prove to be a novel concept of general relevance for controlling the state of matter at the molecular scale, with potential applications to the fields of nanotechnology and biology [47, 48].

Finally, from a methodological point of view, in Sect. 5.5 we have described what is perhaps the only viable approach for an accurate determination of the state-dependent rotational dynamics of dye molecules whenever excited-state absorption cannot be made negligible by a suitable choice of the probing wavelength. Indeed, the excited-state dye cross section for light absorption strongly affects the transient-dichroism signal, making its interpretation ambiguous unless the value of this parameter can be determined independently, as we did. The experimental procedure we have adopted in our investigation may prove to be of general applicability for the study of photoactive guest–host systems.

References

[1] O'Connor, D.V., and Phillips, D. (1984). *Time-Correlated Single Photon Counting*. Academic Press, London.

[2] Lakowicz, J.R. (1999). *Principles of Fluorescence Spectroscopy*, 2nd edition. Kluwer Academic, New York.

[3] Paparo, D., Marrucci, L., Abbate, G., Santamato, E., Kreuzer, M., Lehnert, P., and Vogeler, T. (1997). Molecular-field-enhanced optical Kerr effect in absorbing liquids. *Phys. Rev. Lett.* 78: 38–41.

[4] Muenster, R., Jarasch, M., Zhuang, X., and Shen, Y.R. (1997). Dye-induced enhancement of optical nonlinearity in liquids and liquid crystals. *Phys. Rev. Lett.* 78: 42–45.

[5] Jánossy, I., Lloyd, A.D., and Wherrett, B.S. (1990). Anomalous optical Fréedericksz transition in an absorbing liquid crystal. *Mol. Cryst. Liq. Cryst.* 179: 1–12.

[6] Zwanzig, R. (1978). Rotational friction coefficients of a bumpy cylinder with slipping and sticking boundary conditions. *J. Chem. Phys.* 68: 4325–4326.

[7] Hynes, J.T., Kapral, R., and Weinberg, M. (1978). Molecular rotation and reorientation: Microscopic and hydrodynamic contributions. *J. Chem. Phys.* 69: 2725–2733.

[8] Myers, A.B., Pereira, M.A., Holt, P.L., and Hochstrasser, R.M. (1987). Rotational dynamics of electronically excited aniline in solution from picosecond fluorescence anisotropies. *J. Chem. Phys.* 86: 5146–5155.

[9] Ben-Amotz, D., and Drake, J.M. (1988). The solute size effect in rotational diffusion experiments: A test of microscopic friction theories. *J. Chem. Phys.* 89: 1019–1029.

[10] Alavi, D.S., and Waldeck, D.H. (1991). A test of hydrodynamics in binary solvent systems. Rotational diffusion studies of oxazine 118. *J. Chem. Phys.* 95: 4848–4852.

[11] Hartman, R.S., Alavi, D.S., and Waldeck, D.H. (1991). An experimental test of dielectric friction models using the rotational diffusion of aminoanthraquinones. *J. Chem. Phys.* 95: 7872–7880.

[12] Williams, A.M., Jiang, Y., and Ben-Amotz, D. (1994). Molecular reorientation dynamics and microscopic friction in liquids. *Chem. Phys.* 180: 119–129.

[13] Blanchard, G.J., and Cihal, C.A. (1988). Orientational relaxation dynamics of oxazine 118 and resorufin in the butanols. Valence- and state-dependent solvation effects. *J. Phys. Chem.* 92: 5950–5954.

[14] Alavi, D.S., Hartman, R.S., and Waldeck, D.H. (1991). The influence of wave vector dependent dielectric properties on rotational friction. Rotational diffusion of phenoxazine dyes. *J. Chem. Phys.* 95: 6770–6783.

[15] Blanchard, G.J. (1988). A study of the state-dependent reorientation dynamics of oxazine 725 in primary normal aliphatic alcohols. *J. Phys. Chem.* 92: 6303–6307.

[16] Paparo, D., Manzo, C., Marrucci, L., and Kreuzer, M. (2002). Large deuterium isotope effect in the rotational diffusion of anthraquinone dyes in liquid solution. *J. Chem. Phys.* 117: 2187–2191.

[17] Shen, Y.R. (1984). *The Principles of Nonlinear Optics*. John Wiley & Sons, New York.

[18] Jánossy, I., and Benkler, E. (2003). Light-induced dichroism and birefringence in dye-doped glycerin. *Europhys. Lett.* 62: 698–704.

[19] Kreuzer, M., Marrucci, L., and Paparo, D. (2000). Light-induced modification of kinetic molecular properties: Enhancement of optical Kerr

effect in absorbing liquids, photoinduced torque and molecular motors in dye-doped nematics. *J. Nonlinear Opt. Phys.* 9: 157–182.

[20] Marrucci, L., and Paparo, D. (1997). Photoinduced molecular reorientation of absorbing liquid crystals. *Phys. Rev. E* 56: 1765–1772.

[21] Marrucci, L., Paparo, D., Maddalena, P., Massera, E., Prudnikova, E., and Santamato, E. (1997). Role of guest–host intermolecular forces in photoinduced reorientation of dyed liquid crystals. *J. Chem. Phys.* 107: 9783–9793.

[22] Marrucci, L., Paparo, D., Vetrano, M.R., Colicchio, M., Santamato, E., and Viscardi, G. (2000). Role of dye structure in photoinduced reorientation of dye-doped liquid crystals. *J. Chem. Phys.* 113: 10361–10366.

[23] Kreuzer, M., Hanisch, F., Eidenschink, R., Paparo, D., and Marrucci, L. (2002). Large deuterium isotope effect in the optical nonlinearity of dye-doped liquid crystals. *Phys. Rev. Lett.* 88: 013902(1–4).

[24] Doi, M., and Edwards, S.F. (1986). *The Theory of Polymer Dynamics.* Oxford University Press, Oxford.

[25] Marrucci, L., Paparo, D., Abbate, G., Santamato, E., Kreuzer, M., Lehnert, P., and Vogeler, T. (1997). Enhanced optical nonlinearity by photoinduced molecular orientation in absorbing liquids. *Phys. Rev. A* 58: 4926–4936.

[26] Manzo, C., Paparo, D., Lettieri, S., and Marrucci, L. (2004). Fluorescence-based investigation of the Jánossy effect anomalous wavelength dependence. *Mol. Cryst. Liq. Cryst.* 421:145–155.

[27] Paparo, D., Maddalena, P., Abbate, G., Santamato, E., and Jánossy, I. (1994). Wavelength dependence of optical reorientation in dye-doped nematics. *Mol. Cryst. Liq. Cryst.* 251: 73–84.

[28] Kósa, T., and Jánossy, I. (1995). Anomalous wavelength dependence of the dye-induced optical reorientation in nematic liquid-crystals. *Opt. Lett.* 20: 1230–1232.

[29] Jones, G., Feng, Z., and Bergmark, W.R. (1994). Photophysical properties of (dimethylamino)anthraquinone: Radiationless transitions in solvent and polyelectrolyte media. *J. Phys. Chem.* 98: 4511–4516.

[30] Kreuzer, M., Benkler, E., Paparo, D., Casillo, G., and Marrucci, L. (2003). Molecular reorientation by photoinduced modulation of rotational mobility. *Phys. Rev. E* 68: 011701(1–5).

[31] Inoue, H., Hoshi, T., Yoshino, J., and Tanizaki, Y. (1972). The polarized absorption spectra of some α-aminoanthraquinones. *Bull. Chem. Soc. Japan* 45: 1018–1021.

[32] Petke, J.D., Butler, P., and Maggiora, G.M. (1985). *Ab initio* quantum-mechanical characterization of the electronic states of anthraquinone,

quinizarine, and 1,4-diamino anthraquinone. *Int. J. Quantum Chem.* 27: 71–87.

[33] Bisht, P.B., Joshi, G.C., and Tripathi, H.B. (1995). Excited state hydrogen bonding of the 2-naphthol-triethylamine system in 1,4-dioxane. *Chem. Phys. Lett.* 237: 356–360.

[34] Richtol, H.H., Fitch, B.R., Triplet state acidity constants for hydroxy and amino substituted anthraquinones and related compounds, *Anal. Chem.* 46: 1860–1863.

[35] Scheiner, S., and Cŭma, M. (1996). Relative stability of hydrogen and deuterium bonds. *J. Am. Chem. Soc.* 118: 1511–1521.

[36] Guillot, B., and Guissani, Y. (1998). Quantum effects in simulated water by Feynman–Hibbs approach. *J. Chem. Phys.* 108: 10162–10174.

[37] Cho, C.H., Urquidi, J., Singh, S., Robinson, G.W. (1999). Thermal offset viscosities of liquid H_2O, D_2O, and T_2O. *J. Phys. Chem. B* 103: 1991–1994.

[38] Itzhaki, L.S., and Evans, P.A. (1996). Solvent isotope effects on the re-folding kinetics of hen egg-white lysozyme. *Protein Sci.* 5: 140–146.

[39] Eich, M., Wendorff, J.H., Peck,B., and Ringsdorf, H. (1988). Reversible digital and holographic optical storage in polymeric liquid crystals. *Makromol. Chem. Rapid. Commun.* 8: 59–63.

[40] Ikeda, T., and Tsutsumi, O. (1995). Optical switching and image storage by means of azobenzene liquid-crystal films. *Science* 268:1873–1875.

[41] Berg, R.H., Hvilsted, S., and Ramanujam, P.S. (1996). Peptide oligomers for holographic data storage. *Nature* 383: 505–508.

[42] Gibbons, W.M., Shannon, P.J., Sun, S.-T., and Swetlin, B.J. (1991). Surface-mediated alignment of nematic liquid crystals with polarized laser light. *Nature* 351: 49–50.

[43] Schadt, M., Seiberle, H., and Schuster, A. (1996). Optical patterning of multidomain liquid-crystal displays with wide viewing angles. *Nature* 381: 212–215.

[44] Sekkat, Z., and Dumont, M. (1992). Photoassisted poling of azo dye polymeric films at room temperature. *Appl. Phys. B* 54: 486–489.

[45] Ikeda, T. (2003). Photomodulation of liquid crystal orientations for photonic applications. *J. Mater. Chem.* 13: 2037–2057.

[46] Marrucci, L., Paparo, D., and Kreuzer, M. (2001). Fluctuating-friction molecular motors. *J. Phys.: Condens. Matter* 13: 10371–10382.

[47] Hugel, T., Holland, N.B., Cattani, A., Moroder, L., Seitz, M., and Gaub, H.E. (2002). Single-molecule optomechanical cycle. *Science* 296: 1103–1106.

[48] van Delden, R.A., Koumura, N., Harada, N., and Feringa, B.L. (2002). Unidirectional rotary motion in a liquid crystalline environment: Color tuning by a molecular motor. *Proc. Natl. Acad. Sci. USA* 99: 4945–4949.

Topic Index

Printed in the United States of America.